Telecourse Student Guide

for

UNIVERSE
The Infinite Frontier

Fifth Edition

Stephen P. Lattanzio

Joel M. Levine

Valerie Lynch Lee

THOMSON

BROOKS/COLE

Australia • Canada • Mexico • Singapore • Spain • United Kingdom • United States

Coast Community College District

William M. Vega, Chancellor

Ding-Jo Currie, President, Coastline Community College

Dan C. Jones, Administrative Dean, Office of Instructional Systems Development

Laurie R. Melby, Director of Production

Judy Garvey, Publications Supervisor

Wendy Sacket, Senior Publications Assistant

Thien Vu, Publications Assistant

Content Advisors: Stephen P. Lattanzio, Professor of Astronomy, Orange Coast College
Joel M. Levine, Professor, Physics and Astronomy, Orange Coast College

For more information about our products, contact us at:
Thomson Learning Academic Resource Center
1-800-423-0563

ISBN 0-534-39284-9

Printed in the United States of America.
1   2 3 4 5 6 7 07 06 05 04 03

Asia
Thomson Learning
5 Shenton Way #01-01
UIC Building
Singapore 068808

Australia/New Zealand
Thomson Learning
102 Dodds Street
Southbank, Victoria 3006
Australia

Canada
Nelson
1120 Birchmount Road
Toronto, Ontario M1K 5G4
Canada

Europe/Middle East/Africa
Thomson Learning
High Holborn House
50/51 Bedford Row
London WC1R 4LR
United Kingdom

Preface

To the Student

Welcome to the telecourse *UNIVERSE: The Infinite Frontier*, an introductory course in astronomy. Whether you are planning a career in science, taking the course as part of an academic course for college credit, or taking it just for pleasure, we believe that you will find this telecourse interesting, entertaining, enriching, and inspiring. *UNIVERSE: The Infinite Frontier* brings you the latest astronomical discoveries and astrophysical theories. The course covers four major areas: Exploring the Sky, The Stars, The Universe of Galaxies, and Planets in Perspective. In this course you will have a breathtaking view of nature, from particles smaller than the atom to giant superclusters of galaxies. As an introductory astronomy course, *UNIVERSE: The Infinite Frontier* goes beyond describing the contents of outer space. It examines the scientific method and considers both its promises and limitations. It answers many of our questions and poses new ones time after time, so that we are continually probing into the innermost secrets of time and space.

Course Themes

The designer, academic advisors, and producers of *UNIVERSE: The Infinite Frontier* have developed this course around eight principal themes:

1. The universe is dynamic and continually evolving.

2. The universality of physical laws discovered on Earth allows us to analyze and draw conclusions about celestial phenomena that can only be studied at great distances.

3. Scientific conclusions must be based on an exacting comparison of hypotheses to evidence obtained from observations and experimental data.

4. The universe, because it has objects and environments that cannot be duplicated on Earth, is a unique laboratory for testing scientific hypotheses.

5. Because astronomy is a human endeavor, it is subject to both the limitations and enhancements of personal relationships, biases, inspiration, and creativity.

6. Most astronomical knowledge accumulates incrementally, with each new piece of knowledge providing a potential foundation for further understanding.

7. Observational and computational technologies play critical roles in shaping our understanding of the universe.

8. In addition to scientific value, astronomy has practical and philosophical value because humans are participants in, as well as observers of, the universe.

Course Components

Video Programs

As with most college courses, *UNIVERSE: The Infinite Frontier* has a textbook, a study guide, assignments, and tests. In addition, *UNIVERSE: The Infinite Frontier* has as a special learning element that traditional classroom-based college courses do not have: a companion half-hour video program for each of the 26 lessons in the course.

The 26 video programs for *UNIVERSE: The Infinite Frontier* feature leading practitioners, theoreticians, and academics in the fields of astronomy, planetary science, and astrophysics, who describe and explain celestial objects and events. The unique capabilities of the video will introduce you to the night sky, the life cycle of the stars, the universe of galaxies and the history of the universe, and the origin, characteristics, and evolution of the solar system.

The video programs will take you inside the data rooms of leading observatories. You will see the latest images from NASA and the Jet Propulsion Laboratories and from Earth-based telescopes, space observatories, and the Hubble Space Telescope. A special feature of the video programs is scientifically accurate three-dimensional animation and computer graphics to illustrate complex concepts in astronomy.

Textbook

The textbook for this telecourse is *Horizons: Exploring the Universe*, eighth edition, by Michael A. Seeds (Brooks/Cole Publishing Company, 2004). The content of this textbook parallels the content of the video programs, and textbook pages are referenced in reading assignments, learning objectives, and answer keys in this guide.

Telecourse Student Guide

This telecourse student guide is your road map through *UNIVERSE: The Infinite Frontier*. It is a starting point for each lesson because it contains step-by-step assignments for reading, viewing, and doing related activities, overviews of each lesson's content and the accompanying video program, and a complete array of learning activities to help you master the learning objectives for the lesson.

Each lesson in this guide has the following components:

Assignments: Detailed instructions on activities and reading assignments to be completed before and after viewing each video program.

Overview: An introduction to the main topics covered in the textbook and video program.

Learning Objectives: Statements of what you should learn from reading the textbook assignments, completing the activities in this guide, and viewing the video programs.

Viewing Notes: A capsule description of the video program for each lesson plus questions for you to consider while watching the program. The Viewing Notes are *not* a substitute for actually watching the video program.

Review Activities: Matching and Completion exercises to help you review and reinforce your understanding of important terms and concepts.

Self-Test: A brief multiple-choice quiz that allows you to test your understanding of the material in the lesson.

Using What You've Learned: Many of the lessons include suggestions for additional activities to enhance your understanding and knowledge of astronomy. (Your instructor will indicate how you are to incorporate these exercises into your studies and which ones to do, and he or she may designate them as assignments for extra credit.)

Background Notes: In addition to the sections described above, many lessons in this student guide have accompanying Background Notes, which are assigned in addition to the textbook material. The Background Notes contain information that supplements material presented in the video programs and textbook. Your careful study of the Background Notes is essential to your achieving the Learning Objectives stated for the lesson. The Background Notes for any given unit of this guide appear in a separate section following the last lesson in that unit, as shown in the table of contents.

Answer Key: This section, at the end of the guide, provides answers for the Matching and Completion items and for the Self-Test questions. Check your answers after you have completed each activity. If you have any incorrect answers, review the material.

How to Take a Telecourse

If you are new to college courses in general, and to telecourses in particular, you can perhaps profit from a few suggestions offered by students who have successfully completed other telecourses.

Telecourses are designed for busy people—people with full-time jobs or family obligations—who want to take a course at home, fitting the study into their own personal schedules. To complete a telecourse successfully, you will need to plan in advance how to schedule your viewing, reading, and study. Buy the textbook, telecourse student guide, and any other materials required by your instructor before the course begins and look them over; familiarize yourself with any materials supplied by your college and estimate how much time it will take you to complete special tests and assignments for each lesson. Write the dates of midterms, finals, review sessions, and special projects on your calendar so that you can plan to have extra time to prepare for them. You may find it enjoyable and instructive to watch the programs with other people, but save the talking and discussion until after the programs so that you won't miss important information. After the program, take a few minutes to write a brief summary of what you have seen and the meaning of key concepts and terms and to answer the questions listed at the end of the Viewing Notes for each lesson.

The following suggestions about *UNIVERSE: The Infinite Frontier* have been compiled from students who have satisfactorily completed telecourses.

- *Do* buy both the textbook and the telecourse student guide for *UNIVERSE: The Infinite Frontier* or arrange to share copies with a friend. Do *not* try to get through this course without these books.

- *Do* watch each of the video programs. To pass the examinations, you will need to read and study the textbook and to view the programs. If you have a videocassette recorder, tape the programs for later review.

- *Do* keep up with your work for this course every week. Even if you do not have any class sessions on campus or any assignments to turn in, you should read the textbook and do the assignments in the telecourse student guide, as well as watch the video programs. Set aside viewing, reading, and study time each week and stick to your schedule.

- *Do* get in touch with the faculty member who is in charge of *UNIVERSE: The Infinite Frontier* at your college or university. The instructor can answer any questions you have about the material covered in the course. Your faculty member can also help you catch up if you are behind, advise you about additional assignments, discuss the type of test questions you can expect, and tell you where you might be able to watch programs you have missed or wish to review.

- *Do* complete all the Review Activities and Self-Tests provided in this guide. These will help you master the Learning Objectives and prepare for formal examinations.

- If you miss a program or fall behind in your study schedule, don't give up. Many television stations repeat broadcasts of the programs later in the week or on weekends. Your college might have videocassette copies of programs available in the campus library or media center. And *do* call on your telecourse instructor or manager to help if you have problems of any kind. This person is assigned specifically to help you succeed in *UNIVERSE: The Infinite Frontier*.

Acknowledgments

Producing the *UNIVERSE: The Infinite Frontier* telecourse was a complex team effort by many skilled people. Several of those persons responsible for this course are listed on the copyright page of this book.

In addition to those individuals, appreciation is expressed for the contributions of a number of academic advisors to the telecourse. First, Michael A. Seeds, Ph.D., professor of astronomy at Franklin and Marshall College and author of the textbook, served as a technical advisor to the series and helped in numerous ways throughout production of the video programs.

Production of *UNIVERSE: The Infinite Frontier* was made possible with grants from several educational institutions. Each of them provided a faculty advisor who helped to formulate the overall course outline. These reviewers and the institutions they represented are Michael J. Matkovich, M.S., Oakton Community College (Northern Illinois Learning Resources Consortium); William McCord, M.S., Valencia Community College (Florida State Department of Education); and David Pierce, Ph.D., El Camino College (INTELECOM). These reviewers were joined by six other astronomers to form a national academic advisory committee for the telecourse. The committee members proposed the telecourse themes and objectives, reviewed the video programs as they were developed, and provided innumerable valuable suggestions on content and resources for the video programs. The members of the advisory committee are listed below.

NATIONAL ACADEMIC ADVISORY COMMITTEE MEMBERS

George Jacoby, Ph.D., Kitt Peak National Observatory

Stephen P. Lattanzio, M.A., Orange Coast College

Joel M. Levine, M.A., Orange Coast College

William McCord, M.S., Valencia Community College

Michael Seeds, Ph.D., Franklin & Marshall College

Benjamin Peery, Ph.D., Howard University

David Pierce, Ph.D., El Camino College

Michael J. Matkovich, M.S., Oakton Community College

Harding Eugene Smith, Jr., Ph.D., University of California, San Diego

Sally Stephens, Ph.D., Astronomical Society of the Pacific

Appreciation is also expressed to Brooks/Cole Publishing Company for its support of the telecourse production.

The 26 programs of *UNIVERSE: The Infinite Frontier* were produced in the studios of KOCE-TV. An affiliate of the Public Broadcasting Service (PBS), KOCE-TV is owned and operated by the Coast Community College District. This student guide and other materials for this course were developed for publication by the Office of Instructional Systems Development at Coastline Community College (Fountain Valley, California), a member of the Coast Community College District.

Contents

The Scale of the Cosmos

Assignments

For the most effective study of this lesson, we suggest that you complete the assignments in the sequence listed below.

Before Viewing the Video Program

- Read the Overview and Learning Objectives for this lesson. Use the Learning Objectives to guide your reading, viewing, and thinking.
- Read Chapter 1, "The Scale of the Cosmos," pages 1–8, in the *Horizons* textbook.
- Study Appendix A, "Units and Astronomical Data," in the textbook. Knowledge of Tables A-1 through A-6 is essential to understanding much of the content of this course. Also familiarize yourself with the information in Tables A-7 through A-9.
- Read Background Notes 1, "Scientific Method," following Lesson 6 in this guide.
- Read the Viewing Notes in this lesson.

View "The Scale of the Cosmos" Video Program

After Viewing the Video Program

- Briefly note your answers to questions listed at the end of the Viewing Notes.
- Review all reading assignments for this lesson, especially the Chapter 1 summary on page 8 in *Horizons,* Background Notes 1 for Unit I, and the Viewing Notes in this lesson.
- Write brief answers to the review questions at the end of Chapter 1 in *Horizons* to be certain you understand the text material.

- Complete the Review Activities in this guide to reinforce your understanding of important terms and concepts. Check your answers with the Answer Key and review when necessary.

- Take the Self-Test in this guide to measure your achievement of the Objectives. Check your answers with the Answer Key and review when necessary.

- Complete the Using What You've Learned activities and any other activities and projects assigned by your instructor.

Overview

In a sense, astronomy is the oldest of sciences because humankind has always studied the stars. For early people on our planet, the sky and the objects moving through it were their clock, their calendar, and their compass. The celestial objects and their associated motions ordered people's lives, much as they still order our lives today. Our days are defined by the rotation of our planet; our years by the orbit of our planet around the sun. All our calendars have been based on the cycles of either the moon or the sun. Our directions are identified by the path of the sun through our sky; and the sun and the stars serve as guideposts in our travels.

The science of astronomy as we think of it today began with the ancient Greeks more than 2,000 years ago. They were the first to observe the sky systematically and to offer explanations for the movement of celestial objects based on reason and logic rather than on myth and the supernatural. But from the time of the Greeks until the invention of the telescope in 1608, astronomical observations were limited by what the naked eye could see. Since the seventeenth century, however, the story of astronomy has been one of studying a frontier that expanded with each advance in telescopes and technology. As astronomers saw more, they realized that the universe was larger, more diverse, and older. They learned that stars and clouds of gas and light were sprinkled throughout the universe. Each discovery challenged their assumptions about the universe.

Although astronomers have observed the sky for centuries, it wasn't until the middle of the nineteenth century that they first measured distances to stars. As telescopes became more powerful and measurements more precise, they learned that the universe was far larger than they had imagined: stars were trillions or thousands of trillions of miles away. A new vocabulary evolved to describe these distances. The light-year was a measure of distance and a term that revealed we could not ignore the element of time in our study of the universe. The early part of the twentieth century saw more advances in our understanding of the universe. Astronomers learned how stars shine and measured the distances to "spiral nebulae." Discovery of these spiral nebulae, now known as galaxies, revealed a universe that stretched millions of trillions of miles, if not farther, and pushed the boundaries of astronomical exploration.

Throughout the twentieth century, astronomers not only pushed the frontier outward, they also began to explore it in different ways. They learned that the universe reveals itself not only in visible light but also in the energies (light) of the infrared, radio, gamma, ultraviolet, and X ray. Astronomers placed telescopes at the tops of mountains and in orbit around Earth. Rockets put satellites into space, where they can orbit for years above Earth's atmosphere and send back data about the energies that never reach the surface of Earth. We have sent spacecraft to land humans on the moon and to explore close up all but one of the planets in our solar system. Spacecraft have orbited a few of the planets and landed on some of them. Astronomy, the oldest of sciences, has become humankind's newest science, with technology leading astronomers, almost daily, to discoveries that challenge our understanding.

The programs and lessons in this telecourse take you on an exciting and fascinating journey to familiarize you with the infinite frontier we call the universe. On this journey, you will learn about things you may have never thought possible. You will stretch your imagination to visualize objects that appear only as points of light in a telescope, but, in actuality, may consist of billions of stars or reveal the most violent events in the universe. You will learn about the laws that govern the universe and some of the hypotheses and theories that may change our present knowledge. You will also learn about the people who practice astronomy and why they find it such a fascinating and important human endeavor. We are the only species on Earth that can wonder about the universe, and we are the only species on Earth that can explore the infinite frontier.

Learning Objectives

When you have completed all assignments for this lesson, you should be able to:

1. Identify the major structures of the universe and order them according to size and distance from Earth. *HORIZONS* TEXTBOOK PAGES 4–7; VIDEO PROGRAM.

2. Understand the relative distances of objects in the universe. *HORIZONS* TEXTBOOK PAGES 3–7; VIDEO PROGRAM.

3. Understand and use scientific notation. *HORIZONS* TEXTBOOK PAGE 4 AND APPENDIX A, PAGES 455–456.

4. Define and explain the terms light-year and astronomical unit. *HORIZONS* TEXTBOOK PAGES 4–5; VIDEO PROGRAM.

5. Explain what is meant by the scientific method and identify its principal components. BACKGROUND NOTES 1 in this telecourse student guide.

Viewing Notes

The first video program of *UNIVERSE: The Infinite Frontier* provides a preview of the celestial objects and events you will learn about in the course. Also, you will gain an understanding of the vast distances and the scale of the universe.

"The Scale of the Cosmos" opens with astronomers explaining why they study the stars and planets and about the excitement of the field and its importance to human knowledge. The program then goes on a visual tour of the universe, starting with the sun and moving to the very edge of the known universe. This journey, illustrated with stunning computer animation and photographs from NASA, the Jet Propulsion Laboratory (JPL), and astronomical observatories, introduces the stars, galaxies, and planets that you will learn more about later in the telecourse. As the narration describes the journey out from the sun, you will learn about the immense scale of the universe and how we define distances in terms of the astronomical unit and light-year.

In the segment entitled "Vast," astronomers explain the enormous distances in the universe, providing several analogies with Earthbound objects and distances. Computer graphics reinforce the concept of the universe's immense scales of time and distance. In addition to being incredibly vast and old, the universe is richly populated with trillions of constantly evolving objects. Photographs and computer animation show some of these fascinating celestial objects: galaxies, white dwarfs, red giants, supernovae, pulsars, black holes, and quasars.

This introduction to astronomy concludes by presenting the basic assumption on which the science of astronomy is based: that the physical laws we observe on Earth are the same everywhere—that the laws of nature are, indeed, universal. Astronomers describe the unique challenges of the science and how they study the stars and apply these physical laws. As Michael Seeds, author of the textbook for this course, says in this segment, astronomers "try to figure out what the rules are that describe how the universe works." Finally, several astronomers who will appear in later video programs pose some of the fundamental questions that astronomy seeks to answer about the universe. The search for answers to those questions is the true motivation for studying the infinite frontier.

As you watch the video program, consider the following questions:

1. What is the astronomical unit and when is it used most conveniently?

2. What is the light-year and when is it used most conveniently?

3. How is distance related to the past history of the universe?

4. What do we mean when we say that the "laws of physics are universal"?

5. What questions do astronomers attempt to answer?

Review Activities

Matching

Match the terms listed below with the definitions that follow. Check your answers with the Answer Key and review any terms you missed.

_____ 1. solar system   _____ 4. light-year

_____ 2. scientific notation   _____ 5. Milky Way Galaxy

_____ 3. astronomical unit   _____ 6. spiral arm

a. the system of recording very large or very small numbers by using powers of 10

b. the sun and its planets, asteroids, comets, and other minor objects

c. area of our galaxy where stars are born in great clouds of gas and dust

d. a unit of distance; the distance light travels in one year

e. the large system of billions of stars in which our sun and solar system are located

f. average distance from Earth to the sun

Completion

Fill each blank with the most appropriate term from the list for that paragraph. A term may be used once, more than once, or not at all. Check your answers with the Answer Key and review when necessary. If a question requires two or more answers in succession, they may be in any order, otherwise unless indicated.

1. Our sun is a _____ and belongs to a galaxy called the _____. The sun is just one of billions of _____ that make up the galaxy. Our galaxy is one of about two dozen galaxies that make up a _____. This small group belongs to a much larger group of galaxies, known as a _____. In turn, these larger groups of galaxies are linked to form _____, which may be the largest structures in the _____. In addition to containing so many stars, most galaxies are located millions, if not billions of _____ away. When astronomers study these objects, they must consider these great distances because it takes _____ for the light from these distant objects to reach Earth. Thus, when we look at these objects, we do not see them as they presently are, but as they were, _____ or _____ of years ago.

billions	millions
cluster	star
filaments	stars
galaxies	supercluster
galaxy	thousands
light-years	time
Milky Way	universe

2. Our planet, _____, orbits a medium-sized _____ that we call the _____. When astronomers refer to distances in the solar system, they do so in terms of the distance between _____ and the _____. This distance is called an _____. For example, the distance from the _____ to Mars is 1.5 _____, and the distance from the _____ to Jupiter is 5.2 _____. These distances also imply that it takes a long _____ for light from the sun to reach objects in our solar system. Light from the sun takes about _____ to reach Earth and about _____ to reach Neptune.

In comparison with distances within the solar system, distances to the stars outside our solar system are _____ of times greater. Furthermore, the time it takes light to travel to the stars is on the order of _____. To describe these distances, astronomers use a larger distance unit called a _____, which is defined as the _____ light travels in one _____. This measurement is equal to about _____. The star nearest to our sun is _____. It is at a distance of 4.2 _____. This means that light, leaving this star today, will take 4.2 _____ to reach Earth.

4 hours	astronomical units	Proxima Centauri
4 minutes	distance	Saturn
8 hours	Earth	star
8 minutes	Earth-sun	sun
33,000 astronomical units	hours	thousands
	light-year	time
63,000 astronomical units	light-years	year
	minutes	years
astronomical unit		

Self-Test

Select the one best answer for each question. Check your answers with the Answer Key and review when necessary.

1. Of the following objects, the smallest one is

   a. a star.

   b. a galaxy.

   c. a planet.

   d. the universe.

2. Which of the following objects is the most distant from Earth?

   a. the Andromeda Galaxy

   b. Pluto

   c. the Milky Way Galaxy

   d. the sun

3. The distance between the sun and the nearest star is about

   a. 260 AU.

   b. 260,000 AU.

   c. 2.6 million AU.

   d. 2.6 billion AU.

4. Light takes approximately 8 minutes to travel from the sun to Earth. To reach Neptune, it takes light from the sun about

    a. half as long.

    b. 100 times longer.

    c. 1,000,000 times longer.

    d. 30 times longer.

5. The most convenient method by which to write extremely large or extremely small numbers is called

    a. scientific method.

    b. notation of the scientific.

    c. scientific notation.

    d. numerical analysis.

6. A shorthand way of writing 425,000,000 is

    a. 4.25×10^8.

    b. 425×10^6.

    c. 0.425×10^9.

    d. all of the above.

7. A light-year is

    a. the time it takes Earth to travel around the sun once.

    b. the distance Earth travels in the time it takes light from the sun to reach Earth.

    c. the distance light travels in one year.

    d. equal to 1 million astronomical units (AU).

8. The average Earth-sun distance is referred to as

    a. an astronomical unit.

    b. a light-year.

    c. a kilometer.

    d. a solar mile.

9. The first step in applying the scientific method to the understanding of an event or phenomenon is to

    a. develop a hypothesis.

    b. make observations.

    c. conduct experiments.

    d. develop a theory.

10. When scientists propose a hypothesis to explain a particular event or phenomenon, the hypothesis must be worded in such a way that

    a. it can be proved true.

    b. experiments can be conducted to test the hypothesis.

    c. a theory can evolve from it.

    d. it cannot be contradicted.

Using What You've Learned

According to your own interests or your instructor's assignment, complete one or more of the following activities.

1. While you are taking **UNIVERSE: The Infinite Frontier**, save copies of newspaper and magazine articles on astronomical research and events. Compare the new knowledge gained with material presented in the textbook. Carefully note any new and updated information. Organize the articles into two groups. The first group should contain articles on new and, as of yet, unexplained observations and discoveries; the second group should contain articles that describe observations that support previously proposed theories and hypothesis.

2. While taking this telecourse, develop a bibliography of books and magazines that provide additional information on the topics presented in the video programs and in the textbook. Organize the bibliography around the following major topics: history, instrumentation (telescopes and satellites), stars, galaxies (normal and active), cosmology, the solar system and planets, the search for extraterrestrial life.

3. Construct a scale model of the solar neighborhood, the Milky Way Galaxy, and the Local Group (the small cluster of about 2 dozen galaxies that includes the Milky Way Galaxy). Begin by establishing the scale of 1 light-year per meter. The nearest star, Proxima Centauri, will be a little over 4 meters (4 light-years) away on the model. Calculate the distance to the nearest edge of the Milky Way Galaxy. The galaxy is about 100,000 light-years across, and our solar system is located about two-thirds of the way

out from the center. That puts the sun about 30,000 light-years from the center of the galaxy, or 20,000 light-years from the nearest edge. Determine this distance first in meters and then convert to kilometers (1,000 meters = 1 kilometer). Finally, calculate the scale distance to the edge of the small cluster to which the Milky Way belongs. This cluster is 6 million light-years in diameter. Our own galaxy is about 2.5 million light-years away from another large galaxy called Andromeda. Calculate the distance to the Andromeda Galaxy in our scale (1 meter = 1 light-year). Convert that to kilometers.

If you are really ambitious, acquire a map of your state and, using the scale of the map, place the nearest star, the edge of the Milky Way, and the Andromeda Galaxy on the map. Start at the town in which you live and go outward from there. For example, the edge of the Milky Way is 20,000 light-years from the sun, which, according to our scale, is 20,000 meters from the sun. This distance corresponds to 20 kilometers. Using your map's scale, locate the distance of 20 kilometers from your town. [If you do not understand metric distances, 1 kilometer is equal to 0.6214 mile. Therefore, 20 kilometers is about 12.4 miles. The table on page 455 in the *Horizons* textbook lists metric units and conversion factors.]

Lesson Review

Lesson 1

The Scale of the Cosmos

Please Note: Use this matrix to guide your study and achieve the learning objectives of this lesson. It will also help you to view the video, which defines and demonstrates important concepts and skills as they relate to everyday life.

Learning Objective	Textbook	Telecourse Student Guide	Background Notes
1. Identify the major structures of the universe and order them according to size and distance from Earth.	pp. 4–7	Matching Activities: 1, 5, 6; Completion Activities: 1, 2; Self-Test Questions: 1, 2.	
2. Understand the relative distances of objects in the universe.	pp. 3–7	Completion Activities: 2; Self-Test Questions: 3, 4.	
3. Understand and use scientific notation.	pp. 4–5, 455–456	Matching Activities: 2; Completion Activities: 2; Self-Test Questions: 5, 6.	
4. Define and explain the terms light-year and astronomical unit.	pp. 4–5	Matching Activities: 3, 4; Completion Activities: 1; Self-Test Questions: 7, 8.	
5. Explain what is meant by the scientific method and identify its principal components.		Self-Test Questions: 9, 10.	Background Notes 1: Scientific Method

L E S S O N

2

The Sky

Assignments

For the most effective study of this lesson, we suggest that you complete the assignments in the sequence listed below.

Before Viewing the Video Program

- Read the Overview and Learning Objectives for this lesson. Use the Learning Objectives to guide your reading, viewing, and thinking.

- Read Chapter 2, "The Sky," pages 9–19, and Chapter 3, "Cycles of the Sky," pages 21–25, in the *Horizons* textbook.

- Read Background Notes 2A, "The Roots of Astronomy," and Background Notes 2B, "The Foucault Pendulum," following Lesson 6 in this guide.

- Read the Viewing Notes in this lesson.

View "The Sky" Video Program

After Viewing the Video Program

- Briefly note your answers to questions listed at the end of the Viewing Notes.

- Review all reading assignments for this lesson, especially the Chapter 2 and Chapter 3 summaries on pages 18 and 39 in *Horizons*, Background Notes 2A and 2B for Unit I, and the Viewing Notes in this lesson.

- Respond to the challenges presented by each Critical Inquiry in the reading assignment and write brief answers to the review questions at the end of Chapter 2 and to questions 1–4 at the end of Chapter 3 in *Horizons* to be certain you understand the text material.

- Complete the Review Activities in this guide to reinforce your understanding of important terms and concepts. Check your answers with the Answer Key and review when necessary.

- Take the Self-Test in this guide to measure your achievement of the Objectives. Check your answers with the Answer Key and review when necessary.

- Complete the Using What You've Learned activities and any other activities and projects assigned by your instructor.

Overview

This lesson begins, and the following three lessons continue, the study of the development of modern astronomy. Our current knowledge of the universe was not acquired overnight. The journey toward understanding our universe began thousands of years ago when ancient civilizations first looked upward into the sky and tried to relate what they saw there to their lives on Earth. Early humans may have been bewildered by the sky or even fearful of some phenomena. Some cultures "explained" the sky with complex mythologies.

This lesson starts with the view of the sky as seen by ancient civilizations and traditional peoples. For most ancient peoples, observations of the sky were made for sacred, as well as practical, motives. The sky was part of the supernatural, a place that had been created by supernatural beings and where gods, goddesses, and other revered creatures lived.

Although many civilizations saw sacred meaning in the sky, they also recognized patterns and cycles that existed in the heavens. The same constellations were visible night after night, and different groups of constellations were visible during each season. They made connections between phenomena in the sky and conditions on Earth. They also connected the position of the rising sun and the height of the sun as it moved across the sky with the seasons. Once they recognized these recurring cycles, many ancient peoples developed calendars based on events occurring in the sky.

Each civilization, in its own special way, brought the sky into the daily lives of its people. They named constellations and created mythologies to explain the figures they saw in the sky. Some ancient civilizations, however, made outstanding astronomical observations. The Egyptians, for example, realized that not only did Earth experience a daily and yearly cycle but also a much longer cycle associated with the drift of the North Star. (By the sixteenth century, astronomers recognized that this cycle was due to the precession of Earth's axis.) Other societies built massive structures that may have been used to pinpoint special astronomical sightings. Stonehenge in England, the Big Horn Medicine Wheel in Wyoming, and the pyramids in Egypt are just a few of the ancient observatories mentioned in this lesson.

Although many ancient peoples observed the sky and developed calendars, named constellations, and learned when the seasons would change, they did not develop models or propose theories to explain the phenomena they had observed. They simply believed celestial events and objects were the work of supernatural powers. Thus, these early observations were not based on scientific objectivity and reasoning.

The first break with the traditions of these early astronomers was made by the Greeks, starting around 500 B.C. They not only observed the sky, they used abstract reasoning to develop models and theories to understand the sky and explain what they had seen. One of these Greek astronomers was Hipparchus (second century B.C.), who is featured in this lesson. It was Hipparchus who, after observing the variations in brightness among the stars, devised a stellar classification system. Although he did not explain why the stars varied in brightness, the system he developed (which, with modification, is still in use today) provided a structured framework for studying the stars and objective standards for sorting them into categories. Other Greek astronomers developed the concept of the celestial sphere to explain the perceived motions of objects in the sky. In short, the ancient Greek astronomers were the first persons to superimpose an organized system on the universe.

Learning Objectives

When you have completed all assignments for this lesson, you should be able to:

1. Explain the role of constellations in ancient and modern observations of the sky and name several major constellations. *HORIZONS* TEXTBOOK PAGES 9–13; VIDEO PROGRAM.

2. Explain the concepts of celestial sphere and celestial equator. *HORIZONS* TEXTBOOK PAGES 14–17; VIDEO PROGRAM.

3. Define archaeoastronomy and describe the astronomical observations made by ancient cultures and the astronomical orientations of Stonehenge and other structures built by ancient cultures, including those of the Plains Indians and Maya. BACKGROUND NOTES 2A; VIDEO PROGRAM.

4. Describe the six different brightness classes Hipparchus developed and explain how they became the basis for today's modern stellar magnitude scale. *HORIZONS* TEXTBOOK PAGES 12–13; VIDEO PROGRAM.

5. Describe the tilt of Earth's axis with respect to the ecliptic and explain why this tilt causes the seasons. *HORIZONS* TEXTBOOK PAGES 21–24; VIDEO PROGRAM.

6. Define and identify on a drawing the following: celestial equator, north celestial pole, south celestial pole, ecliptic, vernal equinox, autumnal equinox, summer solstice, winter solstice, sunrise, and sunset. *HORIZONS* TEXTBOOK PAGES 16–18 AND 21–24; VIDEO PROGRAM.

7. Describe the phenomenon of precession and explain its causes. *HORIZONS* TEXTBOOK PAGES 14–15; VIDEO PROGRAM.

8. Describe the evidence that demonstrates that Earth rotates on its axis. BACKGROUND NOTES 2B; VIDEO PROGRAM.

Viewing Notes

"The Sky" explores how different cultures throughout the world over the span of human history have viewed celestial events and objects.

The first half of the video program focuses on ancient sky watchers. Archaeoastronomer Edwin Krupp describes how early cultures used their observations of the sky to organize their lives, and he explains the symbolic meanings they attached to what they saw in the sky. Krupp and other astronomers compare how Navajo and Chinese cultures, for example, named such well-known constellations as the Big Dipper and Scorpius, and they describe the reverence held for the sun by several societies, including those in ancient Egypt and Mesopotamia.

This segment also features footage of two "sacred places"—Stonehenge in England and the Big Horn Medicine Wheel in northern Wyoming. These structures, along with many others throughout the world, were built not only to observe but also to honor celestial objects.

The next segment examines how the ancient Greeks brought a new perspective to the sky and, in doing so, provided the foundation for modern science. Hipparchus' stellar magnitude scale is explained, and computer graphics illustrate the six stellar magnitudes on the scale. The Greeks also developed the first model of the universe, and the video program shows how they used a celestial sphere to explain the observed motions in the sky and developed the concepts of the celestial equator and the ecliptic.

The concluding segment explains the sky as we understand it today and shows how the two major cycles we live with—the day and the year—are determined by the *apparent* motion of the sun. In reality, it is Earth's motions of rotation and orbital revolution around the sun that make our sun and other stars appear to "move" across our sky. Of course, the moon and the planets of our solar system do move across our sky, because they follow their own orbits, independent of Earth's.

Finally, the video program graphically illustrates the changing position of the sun in the sky with time-lapse photography of the sun's path on four different days: the summer solstice, the autumnal equinox, the winter solstice, and the vernal equinox. Computer graphics demonstrate how the tilt of Earth's axis with respect to the sun causes our seasons and how precession causes the location of constellations as seen from Earth to change over time.

As you watch the video program, consider the following questions:

1. What were the reasons ancient civilizations observed the skies and how were these observations incorporated into the daily lives of the people?

2. What are Stonehenge and the Big Horn Medicine Wheel?

3. What is the magnitude scale of Hipparchus?

4. What is the difference between the celestial equator and the ecliptic?

5. How and why do seasons occur? What is meant by the terms equinox and solstice?

6. What is precession and how does it affect our observations of the night sky?

Review Activities

Matching

Match the terms listed below with the definitions that follow. Check your answers with the Answer Key and review any terms you missed.

_____ 1. constellation

_____ 2. asterism

_____ 3. magnitude scale

_____ 4. apparent visual magnitude

_____ 5. celestial sphere

_____ 6. north celestial pole

_____ 7. south celestial pole

_____ 8. ecliptic

_____ 9. vernal equinox

_____ 10. autumnal equinox

_____ 11. summer solstice

_____ 12. winter solstice

_____ 13. precession

_____ 14. perihelion

_____ 15. aphelion

_____ 16. evening star

_____ 17. morning star

_____ 18. rotation

_____ 19. minute of arc

_____ 20. celestial equator

_____ 21. revolution

_____ 22. second of arc

_____ 23. scientific model

_____ 24. zenith

a. apparent path of the sun around the sky

b. point on the celestial sphere where the sun is farthest south

c. a stellar pattern identified by name, usually of mythological gods, people, or animals

d. any planet visible in the sky just before sunrise

e. point on the celestial sphere directly above Earth's South Pole

f. the brightness of a star as seen by human eyes on Earth

g. a named group of stars that is not one of the traditional constellations

h. orbital point at which a planet is farthest from the sun

i. point on the celestial sphere where the sun is at its most northerly point

j. an imaginary sphere of very large radius surrounding Earth and to which the planets, stars, sun, and moon seem to be attached

k. the slow change in direction of Earth's axis of rotation

l. the place on the celestial sphere where the sun crosses the celestial equator moving northward

m. any planet visible in the evening sky just after sunset

n. the point on the celestial sphere directly above Earth's North Pole

o. orbital point at which a planet is closest to the sun

p. the point on the celestial sphere where the sun crosses the celestial equator going southward

q. system for describing the brightness of a star

r. a mental conception, an idea, that helps us think about and understand nature

s. the point in the sky directly overhead

t. an angular unit of measure—each minute has 60 of them

u. spinning movement of an object around its axis (a line passing through the center of the body)

v. imaginary line around the sky directly above Earth's equator

w. an angular unit of measure—each degree has 60 of them

x. orbiting movement around a point outside the orbiting body

Completion

Fill each blank with the most appropriate term from the list for that paragraph. A term may be used once, more than once, or not at all. Check your answers with the Answer Key and review when necessary. If a question requires two or more answers in succession, they may be in any order, unless otherwise indicated.

1. The ancient Greeks and others thought that Earth was surrounded by a great, hollow, crystalline globe with the stars attached to the inside. Although they were incorrect, their model is useful to us when we describe the perceived positions and motions of objects in the sky. This model of the sky is called the _____. If we draw a great circle on this sphere, directly above Earth's equator, we split this crystalline sphere into two equal halves. This circle is called the _____. If we watch the stars on this sphere, they seem to move around a point on the sky directly above Earth's North Pole, a point called the _____. A corresponding point, the _____, can be found in the southern hemisphere directly above Earth's South Pole. If we keep watch on the motion of the planets, the moon, and especially the sun, we notice that these objects move in a very narrow region of the sky. We define the path the sun appears to follow within this region of the sky as the _____.

celestial equator	ecliptic	south celestial pole
celestial sphere	north celestial pole	

2. Seasons occur on Earth because Earth's _____ is _____ with respect to the _____, which causes the _____ and the _____ to cross at two points. These points are referred to as the _____ equinox and _____ equinox. When the sun, in its apparent path around Earth, is at the equinox, the lengths of _____ and _____ are _____. These positions mark the beginnings of the seasons of _____ and _____, respectively. On the longest day of the year, the sun is at the _____ and reaches its _____ point in the sky. On the shortest day of the year, the sun is at the _____ and reaches its _____ point in the sky. These two positions mark the beginnings of the seasons of _____ and _____, respectively.

autumn	equal	summer solstice
autumnal	highest	tilted
axis of rotation	lowest	vernal
celestial equator	night	winter
day	spring	winter solstice
ecliptic	summer	

3. A system for sorting the stars into classes was first developed by the Greek astronomer _____. He sorted the stars in the sky into _____ classes according to their _____. This _____ scale assigned the numbers _____ through _____ to the stars he could see. The _____ stars were assigned the _____ numbers. This system was somewhat confusing, because it made the scale seem upside down. In other words, the brightest stars were _____ class stars and the faintest stars were _____ class stars. Modern astronomers have adopted this scale and modified it slightly. The brightness of some stars was recalculated and determined to be brighter than the original scale classes. These stars were then assigned _____ numbers to describe their _____. Since the original scale did not include the factor of _____, these _____ are known as _____. This scale tells us only how bright these stars _____.

apparent visual magnitudes	first	negative
appear	five	one
Aristotle	highest	seven
brightest	Hipparchus	seventh
brightness	largest	six
faintest	lowest	sixth
fifth	magnitude	
	magnitudes	

Self-Test

Select the one best answer for each question. Check your answers with the Answer Key and review when necessary.

1. Astronomers today use the names of constellations to
   a. refer to groups of stars first identified by the ancient Greeks.
   b. identify asterisms.
   c. identify a specific area of the sky.
   d. refer to only the brightest star in a constellation.

2. The principal reason ancient civilizations gave names to groups of stars in the sky was probably to
   a. predict the seasons.
   b. provide a common reference among different cultures.
   c. develop a calendar.
   d. honor gods, heroes, and animals.

3. When the cosmos is described in terms of the celestial sphere, Earth is located
   a. at the center of the sphere.
   b. one-third from the edge of the sphere.
   c. beyond the limits of the sphere and in orbit around it.
   d. equidistant from the edge of the sphere as the sun.

4. The great circle that divides the celestial sphere into two equal hemispheres is called the
   a. ecliptic.
   b. celestial meridian.
   c. lines of magnitude.
   d. celestial equator.

5. The Navajo Indians described the Big Dipper and some associated stars as the
   a. celestial chariot.
   b. revolving female.
   c. revolving male.
   d. swimming ducks.

6. At Stonehenge, the sun rises over the Heelstone on the day of the
   a. summer equinox.
   b. summer solstice.
   c. winter solstice.
   d. spring equinox.

7. According to the magnitude scale established by Hipparchus, a fifth-magnitude star is
   a. fainter than a fourth-magnitude star.
   b. brighter than a fourth-magnitude star.
   c. the brightest type of star.
   d. the faintest type of star.

8. Stars brighter than a first-magnitude star are identified by
   a. high positive numbers.
   b. negative numbers.
   c. low positive numbers.
   d. letters of the Greek alphabet.

9. The ecliptic is the path the sun appears to follow during the course of a year. It also defines the plane of Earth's orbit. The axis of Earth is tilted with respect to the ecliptic plane by about
   a. 50 degrees.
   b. 12.5 degrees.
   c. 23.5 degrees.
   d. 32.5 degrees.

10. The vernal equinox, the first day of spring, occurs when the sun
    a. reaches its highest daytime position in the northern hemisphere.

    b. reaches its lowest daytime position in the southern hemisphere.

    c. crosses the celestial meridian.

    d. crosses the celestial equator.

11. The winter solstice occurs when the sun is
    a. at its lowest position at noon when viewed by northern hemisphere observers.

    b. near the full moon as viewed by southern hemisphere observers.

    c. crossing the celestial equator.

    d. rising due east and setting due west.

12. The ecliptic and celestial equator cross at an angle of about
    a. 45 degrees.

    b. 32.5 degrees.

    c. 23.5 degrees.

    d. 10.0 degrees.

13. As Earth rotates on its axis and revolves around the sun, its axis slowly wobbles and describes a cone in space. This motion is called
    a. oscillation of the axis.

    b. precession.

    c. retrograde motion.

    d. revolution of the axis.

14. One of the most noticeable consequences of precession (given enough time) is the
    a. change of the pole star.

    b. change of the period of Earth's orbit.

    c. disappearance of certain constellations during the year.

    d. appearance of different stars on the celestial equator during the year.

15. The fact that Earth rotates on its axis can be demonstrated by the observation that

    a. the amplitude of swing of a pendulum decreases with time.

    b. the direction of the plane of swing of a pendulum changes.

    c. a pendulum will rotate at the same rate Earth rotates.

    d. a pendulum will stop swinging after 24 hours.

16. A device that demonstrates that Earth rotates on its axis is a

    a. celestial pendulum.

    b. Newtonian pendulum.

    c. Foucault pendulum.

    d. rotational pendulum.

Using What You've Learned

According to your own interests or your instructor's assignment, complete one or more of the following activities.

1. Observe the rising (or setting) sun once a week during the course of the semester (try to continue this observation for an entire year). On a diagram of the horizon (eastern or western), record the position of the rising or setting sun. Discuss how these observations could lead to the development of a yearly calendar.

2. Place a two- or three-foot-long stick vertically in the ground. Record the length of the shadow once an hour from sunrise to sunset. How would you determine the time of local noon? How could you use this information to determine the longest day of the year, the date of the summer solstice? How could you determine the day of the winter solstice? The Chinese and many other civilizations made these observations thousands of years ago.

3. Select an ancient civilization and conduct research into the astronomy done by that civilization. Be sure to distinguish between their mythology of star names and constellations and their observations of significant astronomical events such as solstices, equinoxes, heliacal rising of particular stars, and eclipses.

Lesson Review
Lesson 2
The Sky

Please Note: Use this matrix to guide your study and achieve the learning objectives of this lesson. It will also help you to view the video, which defines and demonstrates important concepts and skills as they relate to everyday life.

Learning Objective	Textbook	Telecourse Student Guide	Background Notes
1. Explain the role of constellations in ancient and modern observations of the sky and name several major constellations.	pp. 9–13	Matching Activities: 1, 2; Self-Test Questions: 1, 2.	
2. Explain the concepts of celestial sphere and celestial equator.	pp. 14–17	Matching Activities: 5, 16, 17, 18, 19, 20, 22, 23, 24; Completion Activities: 1; Self-Test Questions: 3, 4.	
3. Define archaeoastronomy and describe the astronomical observations made by ancient cultures and the astronomical orientations of Stonehenge and other structures built by ancient cultures, including those of the Plains Indians and Maya.	pp. 455–456	Self-Test Questions: 5, 6.	Background Notes 2A: The Roots of Astronomy
4. Describe the six different brightness classes Hipparchus developed and explain how they became the basis for today's modern stellar magnitude scale.	pp. 12–13	Matching Activities: 3, 4; Completion Activities: 3; Self-Test Questions: 7, 8.	
5. Describe the tilt of Earth's axis with respect to the ecliptic and explain why this tilt causes the seasons.	pp. 21–24	Matching Activities: 8, 21; Completion Activities: 2; Self-Test Questions: 9, 10.	

Learning Objective	Textbook	Telecourse Student Guide	Background Notes
6. Define and identify on a drawing the following: celestial equator, north celestial pole, south celestial pole, ecliptic, vernal equinox, autumnal equinox, summer solstice, winter solstice, sunrise, and sunset.	pp. 16–18, 21–24	Matching Activities: 6, 7, 9, 10, 11, 12, 14, 15; Completion Activities: 2; Self-Test Questions: 11, 12.	
7. Describe the phenomenon of precession and explain its causes.	pp. 14–15	Matching Activities: 13; Self-Test Questions: 13, 14.	
8. Describe the evidence that demonstrates that Earth rotates on its axis.		Self-Test Questions: 15, 16.	Background Notes 2B: The Foucault Pendulum

3

Cycles of the Sky

Assignments

For the most effective study of this lesson, we suggest that you complete the assignments in the sequence listed below.

Before Viewing the Video Program

- Read the Overview and Learning Objectives for this lesson. Use the Learning Objectives to guide your reading, viewing, and thinking.

- Read Chapter 3, "Cycles of the Sky," sections 3-2 and 3-3, pages 25–39, in the *Horizons* textbook.

- Read the Viewing Notes in this lesson.

View the "Cycles of the Sky" Video Program

After Viewing the Video Program

- Briefly note your answers to questions listed at the end of the Viewing Notes.

- Review all reading assignments for this lesson, especially the Chapter 3 summary on page 39 in *Horizons* and the Viewing Notes in this lesson.

- Respond to the challenges presented by each Critical Inquiry in the reading assignment and write brief answers to review questions 5–12 at the end of Chapter 3 in *Horizons* to be certain you understand the text material.

- Complete the Review Activities in this guide to reinforce your understanding of important terms and concepts. Check your answers in the Answer Key and review when necessary.

- Take the Self-Test in this guide to measure your achievement of the Objectives. Check your answers with the Answer Key and review when necessary.

- Complete the Using What You've Learned activities and any other activities and projects assigned by your instructor.

Overview

Even a casual observer soon realizes that changes occur in the sky and on Earth. The moon changes its shape and its position against the background stars. The level of the ocean at the shoreline varies from high to low. The sun on rare occasions darkens during the daytime, and the full moon on occasion turns reddish-orange at night. Tides and eclipses are the most obvious and dramatic manifestations of change in the cycles of the sky. The serious observer will notice that these changes are periodic and, with careful study, predictable. These cycles are the topic of this lesson.

Perhaps the most obvious cycle observed from Earth is the changing shapes or phases of the moon. The cycle occurs so regularly that many cultures use the phases of the moon as a measure of time. It is interesting that one of the most common misconceptions about the origin of the moon's phases is that they are caused by a portion of Earth's shadow falling on the moon. In actuality, the phases of the moon depend on how much and which part of the lunar surface is illuminated by sunlight reflected toward Earth.

The moon is bright, and its changes are easily observed. However, other cycles are more subtle, and not everyone on Earth may be able to observe all the phenomena. Tides are one example. Only persons near a shoreline can observe the changing tides. Although two high and two low tides wash up against shorelines each day, the size of those tides depends on the shape of the ocean floor and coastline at the observer's location. In some places, the differences between high and low tides may be more than 30 feet; in other places, the difference may be only a foot or less.

However interesting the phases of the moon and the cycle of tides may be, it is lunar and solar eclipses that, throughout history, have been the most remarkable and observed celestial events. These events are so dramatic that ancient civilizations made extraordinary efforts to predict the occurrence of eclipses. Even today, astronomers take great effort to predict precisely the time and location of eclipses.

Eclipses, especially total solar eclipses, give astronomers additional opportunities to study features on and near the sun that are normally not easily observed from Earth. Accurate predictions also give nonastronomers the opportunity to prepare themselves for an adventure of a lifetime—the observation of a total solar eclipse. This impressive display by nature is free to everyone who is in the right place at the right time. Although no solar eclipses will be visible over North America until 2008 (Canada), several will occur in the Western hemisphere. If someday you should be fortunate enough to observe a solar eclipse, remember: **Never look directly at the sun without using a scientifically approved filter.** These filters are usually available through optical and telescope shops, local astronomy clubs, and planetariums. Please be careful. The intense light from the sun can irreparably damage your eyes in a very short time.

Until you are fortunate enough to view a total solar eclipse, you will have to rely on personal commentaries to gain some insight into the experiences of astronomers and others who have observed a total solar eclipse. Words such as "awesome," "magnificent," "events beyond my control," and "frozen in place" have been used to describe this event. Although these descriptions seem trite and overused, they are in fact appropriate words to describe one of the most amazing natural events ever observed by humans. These feelings of awe, excitement, and curiosity are at the root of our desire to understand and explore events and phenomena within our universe.

This lesson also examines the astronomical influences on Earth's climate, including such worldwide climate changes as periodic ice ages.

Learning Objectives

When you have completed all assignments for this lesson, you should be able to:

1. Explain, by drawing a diagram, the phases of the moon and name these phases. *HORIZONS* TEXTBOOK PAGES 26, 28–29; VIDEO PROGRAM.

2. Explain the difference between the sidereal and synodic periods of the moon. *HORIZONS* TEXTBOOK PAGE 29; VIDEO PROGRAM.

3. Describe the relationship between the time for the moon to orbit Earth once (sidereal period) and the time for the moon to rotate on its axis once. *HORIZONS* TEXTBOOK PAGE 29; VIDEO PROGRAM.

4. Describe how the moon and the sun produce tides on Earth and explain the effect of tidal forces on Earth's rotation and the orbital motion of the moon. *HORIZONS* TEXTBOOK PAGES 26-27, 30; VIDEO PROGRAM.

5. Explain the necessary conditions for the occurrence of a lunar eclipse. *HORIZONS* TEXTBOOK PAGES 30–31; VIDEO PROGRAM.

6. Explain the necessary conditions for the occurrence of a solar eclipse and the various types of solar eclipses possible. *HORIZONS* TEXTBOOK PAGES 30–37; VIDEO PROGRAM.

7. Describe the astronomical influences on Earth's climate. *HORIZONS* TEXTBOOK PAGES 37–39.

Viewing Notes

"Cycles of the Sky" begins with a detailed description of the phases of the moon, accompanied by computer graphics that highlight the relative positions of Earth, the moon, and the sun during each of the phases. The video program also explains the sidereal and synodic periods of the moon and why they are different.

This knowledge of the relative positions of Earth, the sun, and the moon provides a foundation for the explanation of tides on Earth caused by differences in gravitational pull on opposite sides of Earth.

The segment on lunar eclipses describes how lunar eclipses result when the moon enters Earth's shadow and explains the differences between penumbral and umbral eclipses. An especially interesting section explains why, in a total lunar eclipse, the moon sometimes appears to be a coppery red disc.

The comparatively rare but spectacular phenomenon of a solar eclipse is the next topic covered in the video program. Astronomers explain how the unique orbital and size relationships between Earth, the sun, and the moon make a total solar eclipse possible. Photographs and film of actual solar eclipses illustrate such phenomena as Bailey's beads and the diamond ring effect. The program also explains why eclipses do not occur every month but do occur at predictable intervals, known as the Saros cycle.

In the final segment, three astronomers describe their personal thoughts, feelings, and experiences while observing the total solar eclipse that occurred on July 11, 1991. This segment includes footage of preparations for viewing, the progress of the eclipse itself, the changing quality of light in the area of the total eclipse, the pinhole projections of light through leaves, and the solar corona and prominences visible only during a total solar eclipse.

As you watch the video program, consider the following questions:

1. What causes the phases of the moon?

2. What are the different phases of the moon and what is the appearance of each phase?

3. What is the difference between the sidereal and synodic periods of the moon?

4. What causes the tides and what is meant by differential attraction?

5. What conditions are necessary for a lunar eclipse to occur?

6. What conditions are necessary for a solar eclipse to occur?

7. What major observations can amateurs and professionals make during a total eclipse of the sun?

Review Activities

Matching

Match the terms listed below with the definitions that follow. Check your answers with the Answer Key and review any terms you missed.

_____ 1. sidereal period   _____ 11. corona

_____ 2. synodic period   _____ 12. prominences

_____ 3. spring tide   _____ 13. diamond ring effect

_____ 4. neap tide   _____ 14. apogee

_____ 5. lunar eclipse   _____ 15. annular eclipse

_____ 6. penumbra   _____ 16. nodes

_____ 7. umbra   _____ 17. eclipse seasons

_____ 8. solar eclipse   _____ 18. Saros cycle

_____ 9. photosphere   _____ 19. perigee

_____ 10. chromosphere

a. ocean tide of high amplitude that occurs at full and new moon

b. the event that occurs when the moon passes directly between Earth and the sun, blocking our view of the sun

c. the points where an object's orbit passes through the plane of Earth's orbit

d. eruptions on the solar surface visible during total solar eclipses

e. the point closest to Earth in the orbit of a body circling Earth

f. the faint outer atmosphere composed of low-density, high-temperature gas visible during a total solar eclipse

g. the portion of a shadow that is only partially shaded

h. the time a celestial body takes to turn once on its axis or revolve once around its orbit relative to the stars

i. ocean tide of low amplitude occurring at first- and third-quarter moon

j. the point farthest from Earth in the orbit of a body circling Earth

k. the bright visible surface of the sun

l. an 18-year, 11-day period after which the pattern of lunar and solar eclipses repeats

m. an object's orbital period with respect to the sun

n. the region of a shadow that is totally shaded

o. a solar eclipse in which the photosphere appears around the edge of the moon in a brilliant ring, or annulus

p. during a total solar eclipse, the momentary appearance of a spot of photosphere at the edge of the moon, producing a brilliant glare set in the silvery ring of the corona

q. the darkening of the moon when it moves through Earth's shadow

r. bright gases just above the photosphere of the sun

s. the intervals of time, approximately six months apart, during which eclipses are possible

Completion

Fill each blank with the most appropriate term from the list for that paragraph. A term may be used once, more than once, or not at all. Check your answers with the Answer Key and review when necessary.

1. A total solar eclipse is probably one of the most impressive events in all of nature. This kind of eclipse occurs when the shadow of the _____ falls onto the surface of _____. The darkest part of the lunar shadow is called the _____, and only those individuals within this portion of the moon's shadow see the total eclipse. During the eclipse, the moon blocks out the _____ of the sun. Once totality is reached, the sun's _____, or faint outer atmosphere, becomes visible. Also visible during a solar eclipse are the atmosphere immediately above the sun's visible surface, the _____, and streamers of solar material, known as _____, that stretch thousands of miles into space.

chromosphere	penumbra
corona	photosphere
Earth	prominences
moon	umbra

2. In order for a solar eclipse to occur, the moon must be at _____ phase. For a lunar eclipse to occur, the moon must be at _____ phase. These phases occur every month, yet eclipses do not occur every month because the moon's orbit is _____ with respect to _____ orbit, which causes the moon's orbit to cross _____ orbit twice at points called _____. The moon must be at or near these points, and at the right phase, in order for an eclipse to occur. The periods of time during which eclipses are possible are referred to as _____. These periods occur _____ times a year and are about _____ months apart. Furthermore, since the moon's orbit is _____ and not perfectly circular, the moon may be at its greatest distance from Earth, or _____. As a result, the moon will appear _____ in the sky and not be able to cover the sun completely. The result will be a partial exposure of the _____ of the sun in the form of a _____ or _____. This type of eclipse is called an _____ eclipse.

annular	full	six
annulus	new	smaller
apogee	nodes	sun's
Earth's	perigee	three
eclipse seasons	photosphere	tilted
elliptical	ring	two
four		

Self-Test

Select the one best answer for each question. Check your answers with the Answer Key and review when necessary.

1. When the moon is located in a direct line and between Earth and sun, the phase is called
   a. full moon.
   b. last quarter.
   c. new moon.
   d. first quarter.

2. The angle formed by the sun, Earth, and moon with Earth at the vertex during the third-quarter phase of the moon is
   a. 90°.
   b. 180°.
   c. 45°.
   d. 0°.

3. The sidereal period of the moon is the length of time it takes for the moon to
   a. go through its cycle of eclipses, both lunar and solar.
   b. orbit Earth and return to the same position relative to the background stars.
   c. complete its cycle of phases.
   d. change from new to full phase.

4. The length of time it takes for the moon to complete its cycle of phases is referred to as the
   a. Saros cycle.
   b. sidereal period.
   c. synodic period.
   d. lunar month.

5. The moon rotates on its axis
   a. once every 24 hours.
   b. in the same amount of time it takes for the moon to orbit Earth.
   c. once every 24 days.
   d. in a period of time equal to the cycle of the tidal bulges on Earth.

6. Because its rotation rate is synchronized with its orbital motion, the moon
   a. causes tides on Earth every 12 hours.
   b. can, at times, block out sunlight and produce a solar eclipse.
   c. enters Earth's shadow to produce a lunar eclipse.
   d. keeps the same side pointed toward Earth at all times.

7. The bulge in Earth's oceans that occurs on the side of Earth nearest the moon results from
   a. the moon's gravitational influence being greater on the side of Earth nearest the moon than it is on the center of Earth.
   b. the moon's gravitational influence being weaker on the side of Earth nearest the moon than it is on the center of Earth.
   c. Earth's gravitational influence on the oceans being weaker than the moon's.
   d. Earth's rotation, which creates centrifugal forces.

8. The extreme tides of the month that occur when the sun's and moon's gravitational influence operate together and increase the net force of gravity on the oceans of Earth are referred to as
   a. neap tides.
   b. spring tides.
   c. lunar tides.
   d. solar tides.

9. In order for a total lunar eclipse to occur,
   a. Earth must pass through the penumbra of the moon's shadow.
   b. the moon must pass through the penumbra of Earth's shadow.
   c. the moon must pass through the umbra of Earth's shadow.
   d. Earth must pass through the umbra of the moon's shadow.

10. A lunar eclipse can occur only if the phase of the moon is

    a. new.

    b. last quarter.

    c. waxing crescent.

    d. full.

11. In order to observe a total solar eclipse, an individual must be located completely within the

    a. penumbra of the moon's shadow.

    b. umbra of the moon's shadow.

    c. umbra of Earth's shadow.

    d. penumbra of Earth's shadow.

12. We do not see either a solar or lunar eclipse every month because

    a. the moon's orbit is tilted with respect to Earth's orbital plane.

    b. the full moon does not occur every month.

    c. the orbit of the moon is too elongated and the moon periodically misses Earth's shadow.

    d. it is not visible from the land masses on Earth each month.

13. Twice a year, the line of nodes points toward the sun. When a full or new moon approaches the line of nodes at this time,

    a. nothing happens.

    b. a lunar eclipse is followed by a solar eclipse within two weeks.

    c. a solar eclipse is followed by a lunar eclipse within four weeks.

    d. an eclipse of one kind or another is possible.

14. Because of the elongated nature of the lunar orbit, the moon is farther from Earth at some times than it is at others. If a solar eclipse occurs when the moon is farthest from Earth, the moon does not cover the sun completely, but leaves a ring of sunlight around it. This eclipse is called

    a. an annular eclipse.

    b. a semi-solar eclipse.

    c. a partial eclipse.

    d. a total eclipse.

15. The Milankovitch hypothesis, stating that changes in Earth's climate have an astronomical origin,

   a. has been universally accepted.

   b. has been universally rejected.

   c. has never been taken seriously.

   d. probably contains some truth and remains the subject of debate and testing.

16. Which of the following is **NOT** one of the astronomical influences at work in the Milankovitch hypothesis?

   a. small changes in Earth's orbit

   b. precession

   c. solar eclipses

   d. inclination

Using What You've Learned

According to your own interests or your instructor's assignment, complete one or more of the following activities.

1. Observe the moon for as many nights during an entire cycle as possible. Make note of (a) the rising and setting times, (b) the motion of the moon against the background stars, and (c) the phase of the moon.

2. Photograph the moon each night for as much of the cycle as possible. (Record the dates.) Be sure to use the same lens for all exposures. (You will need only 1/125 exposure time at 400 ISO for most of these photographs. Check your meter to be sure.) Compare the diameter of the moon in each photograph and determine the dates of apogee and perigee. (The moon will appear largest at perigee.)

3. Make a list of the sources that give information concerning the dates, times, and locations for observations of lunar and solar eclipses that will occur during the next year.

4. Research the history of eclipse observations. What mythologies exist because of eclipses? What mythologies exist that attempt to explain the phases of the moon?

5. The following table lists partial and total lunar eclipses that will be visible from the continental United States through 2014. If at all possible, observe a lunar eclipse.

Partial and Total Lunar Eclipses Visible in the Continental United States, 2004–2014

Note: Times are Eastern standard time, and so may differ slightly depending on the geographical location of the observer in the United States and the local time zone. The three times listed are (a) first contact with umbra (partial or total eclipse), (b) maximum eclipse, (c) last contact with umbra. For more details about each eclipse, refer to the following website:
http://sunearth.gsfc.nasa.gov/eclips

DATE AND TIME	TYPE	VISIBILITY
2004, October 27 8:16 P.M. 10:05 P.M. 11:43 P.M.	Total	Beginning visible on the East Coast; end of the eclipse visible in the West.
2005, October 17 6:34 P.M. 7:02 P.M. 7:30 P.M.	Partial	Eastern United States only.
2007, March 3 4:34 P.M. 6:24 P.M. 8:13 P.M.	Total	Middle and end visible in Eastern United States only.
2007, August 28 · 3:51 A.M. 5:37 A.M. 7:22 A.M.	Total	Beginning and middle visible on the East Coast; all of the eclipse visible in Central and Western United States.
2008, February 20 8:47 P.M. 10:29 P.M. 12:11 A.M.	Total	All of the United States.
2010, June 26 5:19 A.M. 6:40 A.M. 8:00 A.M.	Partial	Beginning visible in Central United States; complete eclipse visible in the West.
2010, December 21 1:34 A.M. 3:20 A.M. 5:01 A.M.	Total	All of the United States.
2011, December 10 7:45 A.M. 9:31 A.M. 11:18 A.M.	Total	Beginning visible on the East Coast. End of totality visible in the West.
2012, June 04 4:59 A.M. 6:03 A.M. 7:06 A.M.	Partial	Beginning visible on the East Coast; end visible in the Midwest
2014, April 15 0:57 A.M. 2:45 A.M. 4:33 A.M.	Total	All of the United States except New England. In New England, the moon will set before the eclipse ends.

Lesson Review

Lesson 3

Cycles of the Sky

Please Note: Use this matrix to guide your study and achieve the learning objectives of this lesson. It will also help you to view the video, which defines and demonstrates important concepts and skills as they relate to everyday life.

Learning Objective	Textbook	Telecourse Student Guide	Background Notes
1. Explain, by drawing a diagram, the phases of the moon and name these phases.	pp. 26, 28–29	Self-Test Questions: 1, 2.	
2. Explain the difference between the sidereal and synodic periods of the moon.	p. 29	Matching Activities: 1, 2; Self-Test Questions: 3, 4.	
3. Describe the relationship between the time for the moon to orbit Earth once (sidereal period) and the time for the moon to rotate on its axis once.	p. 29	Self-Test Questions: 5, 6.	
4. Describe how the moon and the sun produce tides on Earth and explain the effect of tidal forces on Earth's rotation and the orbital motion of the moon.	pp. 26–27, 30	Matching Activities: 3, 4; Self-Test Questions: 7, 8.	
5. Explain the necessary conditions for the occurrence of a lunar eclipse.	pp. 30–31	Matching Activities: 5, 6, 7, 16, 17, 18; Completion Activities: 2; Self-Test Questions: 9, 10.	
6. Explain the necessary conditions for the occurrence of a solar eclipse and the various types of solar eclipses possible.	pp. 30–37	Matching Activities: 8, 9, 10, 11, 12, 13, 14, 15, 16, 17, 18, 19; Completion Activities: 1, 2; Self-Test Questions: 11, 12, 13, 14.	

Learning Objective	Textbook	Telecourse Student Guide	Background Notes
7. Describe the astronomical influences on Earth's climate.	pp. 37–39	Self-Test Questions: 15, 16.	

The Origin of Modern Astronomy

Assignments

For the most effective study of this lesson, we suggest that you complete the assignments in the sequence listed below.

Before Viewing the Video Program

- Read the Overview and Learning Objectives for this lesson. Use the Learning Objectives to guide your reading, viewing, and thinking.

- Read Chapter 4, "The Origin of Modern Astronomy," sections 4-1 through 4-5, pages 42–60 in the *Horizons* textbook.

- Read Background Notes 4, "Eratosthenes' Calculation of Earth's Circumference," following Lesson 6 in this guide.

- Read the Viewing Notes in this lesson.

View "The Origin of Modern Astronomy" Video Program

After Viewing the Video Program

- Briefly note your answers to questions listed at the end of the Viewing Notes.

- Review all reading assignments for this lesson, especially the Chapter 4 summary on page 66 in *Horizons,* Background Notes 4 for Unit I, and the Viewing Notes in this lesson.

- Respond to the challenges presented by each Critical Inquiry in the reading assignment and write brief answers to review questions 1–13 at the end of Chapter 4 in *Horizons* to be certain you understand the text material.

- Complete the Review Activities in this guide to reinforce your understanding of important terms and concepts. Check your answers with the Answer Key and review when necessary.

- Take the Self-Test in this guide to measure your achievement of the Objectives. Check your answers with the Answer Key and review when necessary.

- Complete the Using What You've Learned activities and any other activities and projects assigned by your instructor.

Overview

This lesson surveys our understanding of the solar system from the time of the ancient Greeks to the time of Johannes Kepler. A major change in the way humans described nature began with the Greeks more than 2,000 years ago. Before that time, supernatural beings and events were used to explain the universe. In ancient Greece, such thinkers as Pythagoras, Aristotle, Ptolemy, and Eratosthenes began to use scientific reasoning and mathematics to develop models and hypotheses based on observations. In other words, science began to offer explanations based on reality rather than on divine forces. Although many of their hypotheses were incorrect, the contributions of the Greeks mark the beginning of science as we know it today.

Among the ancient Greeks, Pythagoras (c. 580–c. 500 B.C.), a mathematician and philosopher, was the first to develop the idea that Earth is a sphere (possibly based on observations of the round shadow cast on the moon by Earth during eclipses and of how a ship's mast would gradually disappear as the ship sailed over the horizon). In Pythagoras's model of the universe, Earth was surrounded by several crystalline spheres that rotated and to which the moon, sun, planets, and stars were attached.

Aristotle, 150 years after Pythagoras, refined the Pythagorean concept of the universe into a more detailed model, which served as the foundation for Western civilization's understanding of the universe for almost 2,000 years. In Aristotle's model, Earth was a stationary sphere, encircled by more than 50 crystalline spheres carrying celestial bodies. The spheres rotated westward each day, which, according to Aristotle's model, accounted for the movements of the moon, sun, planets, and stars.

To the philosophers and people of that time, Aristotle's model seemed reasonable. To them, as to us, Earth did not feel as if it were moving. Other "evidence" that Earth did not move was provided by the observation that birds and other airborne objects were not left behind as Earth moved beneath them. For the Greeks, Earth simply did not move.

Another ancient Greek astronomer who employed mathematics and scientific reasoning to learn about Earth was Eratosthenes. Around 200 B.C. he calculated the size of Earth. The method he used is described in Background Notes 4, "Eratosthenes' Calculation of Earth's Circumference."

Because explanations of how events occurred have to be supported by actual observations, Aristotle's model was found to be inadequate for understanding some celestial phenomena, such as the occasional apparent backward motions of the planets. Other Greek astronomers modified Aristotle's model until it became very complicated. Then about A.D. 140, some 500 years after Aristotle developed his model, an astronomer named Ptolemy (Claudius Ptolemaeus) created an even more sophisticated and accurate model of the universe. In Ptolemy's model, Earth was at the center of the universe and was surrounded by a perfect heaven in which planets moved at uniform speed in absolutely circular orbits around Earth. Ptolemy's model required dozens of epicycles and deferents, circles upon circles, to account for the observed motions of planets and astronomical cycles.

It was not until the mid-sixteenth century that a major revolution occurred in the way Western civilization looked at the solar system and universe. Nicolaus Copernicus maintained that the sun was at the center of the universe. In his model, the planets and Earth moved around the sun in circular orbits. The evidence he was able to offer to support his hypothesis, however, was not very convincing. It did not predict the positions of planets more accurately than did the Ptolemaic model nor did it explain the parallax shift in the stars as Earth orbited the sun. What eventually made the heliocentric, or Copernican, model so desirable was that it offered a simple, even elegant, view of nature. Copernicus and other supporters of the heliocentric model concluded that nature did things in a way that is pleasing to both the eye and intellect. The task is to explain nature in the simplest way.

Evidence was still required to support the Copernican (heliocentric) model. It was not until Galileo used a telescope to observe the night sky that such evidence was obtained. The phases of Venus and the discovery of satellites around Jupiter were the convincing observations that proved that celestial objects did not orbit Earth. Even these observations, however, were unable to make the Copernican model more accurate. It took the naked-eye observations of Tycho Brahe (before telescopes were invented) and the mathematical genius of Johannes Kepler to create a set of laws of the solar system that correctly predicted the positions of the planets. Kepler's three laws of planetary motion placed the sun not at the center of a circle but at one of the foci of an ellipse and showed the relationships between the speed of a planet, the period of its orbit, and its average distance from the sun.

Science had come a long way from the time of the ancient Greeks to the time of Johannes Kepler. Models of the universe had to be in agreement with observational evidence. Mathematical rules had to be consistent. However, science still lacked fundamental laws to explain not only what happened in the heavens but also events that occurred on Earth. Pieces of a puzzle were beginning to fit together. Science was on the threshold of a major advance. That advance was started by Isaac Newton and continued by Albert Einstein. Their contributions to our understanding of the universe are topics covered in the next lesson.

Learning Objectives

When you have completed all assignments for this lesson, you should be able to:

1. Describe how Eratosthenes measured the circumference of Earth. BACKGROUND NOTES 4.

2. Describe the geocentric models of the solar system developed by Aristotle and Ptolemy. *HORIZONS* TEXTBOOK PAGES 42–46; VIDEO PROGRAM.

3. Describe the motions of the naked-eye planets as observed from Earth and define the term retrograde. *HORIZONS* TEXTBOOK PAGES 44–45; VIDEO PROGRAM.

4. Explain the Copernican model and hypothesis. *HORIZONS* TEXTBOOK PAGES 46–50; VIDEO PROGRAM.

5. Compare and contrast the Ptolemaic (geocentric) model of the solar system with the Copernican (heliocentric) model. *HORIZONS* TEXTBOOK PAGES 43–50.

6. Identify the various instruments used by Tycho Brahe to make his observations and describe the various conclusions he made based on his observations. *HORIZONS* TEXTBOOK PAGES 50–52; VIDEO PROGRAM.

7. Explain Kepler's three laws of planetary motion. *HORIZONS* TEXTBOOK PAGES 52–56; VIDEO PROGRAM.

8. Describe the telescopic observations of the solar system made by Galileo. *HORIZONS* TEXTBOOK PAGES 56–58; VIDEO PROGRAM.

9. Distinguish between paradigm, hypothesis, model, theory, and law. *HORIZONS* TEXTBOOK PAGES 14 (WINDOW ON SCIENCE 2-1), 49, AND 55.

Viewing Notes

This video program surveys the roots of the modern science of astronomy in the contributions of Aristotle, Ptolemy, Copernicus, Galileo, Tycho Brahe, and Kepler.

"The Origin of Modern Astronomy" begins by examining the ideas of Aristotle, who was one of the first Greeks to apply geometric thinking to explain the motions he observed in the sky. Aristotle's ideas were perpetuated and refined by Ptolemy, who developed a more comprehensive model of the universe. The program explains and illustrates how the rather complex Ptolemaic model accounted for the motions of the planets.

Most of the remainder of the program covers the 99 years, from 1543 to 1642, in which the true foundation of modern astronomy was built from the contributions of Copernicus, Tycho Brahe, Kepler, and Galileo.

The video program first features Nicolaus Copernicus. Several astronomers explain the features of his heliocentric model of the universe and compare it to the earlier Ptolemaic model. Computer animation clearly illustrates how the Copernican model explains the retrograde motion of the planets.

Although the Copernican model was the first that correctly placed the sun at the center of the universe (our solar system), the idea was not widely accepted until Galileo's observations showed that Venus orbited the sun and that objects (moons) orbited Jupiter. These observations clearly suggested that Earth orbited the sun. The video program describes his telescopic observations of the moon and planets and explains how they supported the Copernican model.

Both Copernicus and Galileo experienced personal difficulties with the Roman Catholic Church because of their models and observations. The video program describes how their work conflicted with church dogma.

The video program concludes with a description of the work of Tycho Brahe and Johannes Kepler. Tycho's theories, although in opposition to the Copernican model, were based on meticulous observations made with precise instruments. After Tycho's death, Kepler studied Tycho's observations and developed his three laws of planetary motion, which were the first to predict accurately the motions of the planets. (See time line on page 46 for historical context.)

The Beginnings of Modern Astronomy

c. 580 B.C.	Pythagoras born
500 B.C.	Pythagoras dies
384 B.C.	Aristotle born
322 B.C.	Aristotle dies
200 B.C.	Eratosthenes measures circumference of Earth
A.D. 140	Ptolemy develops model of universe
1473	Nicolaus Copernicus born
1507	Copernicus publishes theory that the sun is the center of the universe
1543	Copernicus dies
1546	Tycho Brahe born
1564	Galileo Galilei born
1571	Johannes Kepler born
1573	Tycho Brahe publishes *De Stella Nova*
1600	Kepler starts to work with Tycho Brahe
1601	Tycho Brahe dies
1609	Kepler publishes *Astronomia Nova*
1610	Galileo publishes *Sidereus Nuncius* describing his early telescopic observations
1633	Galileo sentenced to life imprisonment (house arrest)
1642	Galileo dies

As you watch the video program, consider the following questions:

1. How did the Greek approach to science differ from that of other ancient civilizations?

2. What basic questions were the early astronomers trying to answer?

3. What were the components of the Ptolemaic model of the solar system and how did the model account for the backward, or retrograde, motion of the planets?

4. What were the main features of the Copernican model of the solar system and how did it account for the retrograde motion of the planet Mars?

5. What were the advantages and limitations of the Copernican model?

6. How did Galileo's discovery of satellites around Jupiter and the phases of Venus support the heliocentric model of the solar system?

7. Who was Tycho Brahe and what contributions did he make toward our understanding of the solar system?

8. What are Kepler's three laws of planetary motion?

Review Activities

Matching

Match the terms listed below with the definitions that follow. Check your answers with the Answer Key and review any terms you missed.

_____ 1. parallax	_____ 8. heliocentric universe
_____ 2. geocentric universe	_____ 9. hypothesis
_____ 3. uniform circular motion	_____ 10. model
_____ 4 retrograde motion	_____ 11. theory
_____ 5. epicycle	_____ 12. natural law
_____ 6. deferent	_____ 13. ellipse
_____ 7. equant	_____ 14. paradigm

a. a model of the universe with the sun at the center, such as the Copernican universe

b. in the Ptolemaic theory, a small circle in which a planet moves

c. a model of the universe with Earth at the center

d. a closed curve around two points called foci, such that the total distance from one focus to the curve and back to the other focus remains constant

e. a description of a natural phenomenon

f. the apparent change in the observed position of an object due to a change in the location of the observer

g. a conjecture, subject to testing, to explain certain facts

h. the apparent backward (westward) motion of planets as seen against the background stars

i. the belief that objects in the sky move uniformly along perfectly circular orbits

j. a theory that is almost universally accepted as true

k. in the Ptolemaic theory, the point off center in the deferent from which the center of the epicycle appears to move uniformly

1. a network of assumptions and principles providing a framework for formulating theories

    m. a system of assumptions and principles applicable to a wide range of phenomena that have been repeatedly verified

    n. in the Ptolemaic theory, the large circle around Earth along which the center of the epicycle moved

Completion

Fill each blank with the most appropriate term from the list for that paragraph. A term may be used once, more than once, or not at all. Check your answers with the Answer Key and review when necessary.

1. Ptolemy developed what is commonly called the _____ or _____-centered system. In this model of the solar system, Earth remained _____, and the planets, sun, and moon moved in _____orbits around Earth. Because the Greeks believed the heavens were perfect, Ptolemy's model required the planets to move at _____ speed. This movement of the planets is referred to as _____motion.

    However, the planets, at times, do not always appear to move forward. Occasionally, the planets appear to move backward or _____ against the stars. To explain this perceived backward motion, Ptolemy placed the planets on small circles called _____. In order to maintain the notion of _____ speed, Ptolemy placed these small circles on larger circles called _____.

    Because motions of the planets still did not match the observations satisfactorily, Ptolemy adjusted his model by moving Earth _____. Ptolemy's model was very complicated. Many circles upon circles were required to explain why the planets were observed to move at _____ speed. However, this desired motion could be viewed only from a point called the _____, a point that was not on Earth.

circular   heliocentric

constant   off center

deferents   retrograde

Earth   stationary

epicycles   sun

equant   uniform circular

geocentric

2. Johannes Kepler used the observational data of _____ to
    devise _____ laws that described the motions of the planets.
    These laws expanded upon the _____ or _____-
    centered model of Copernicus. Kepler's laws were the first to connect
    observation with mathematical relationships. In other words, they
    described the solar system as it truly is.

    Kepler's first law states that the orbits of the planets are
    _____ with the _____ as one of the foci. The second
    law states that a line from the planet to the sun will sweep out
    _____ areas in _____ times. This law means that
    when the planet is closest to the sun, or at _____, the planet
    travels _____ in its orbit. When the planet is farthest from the
    sun, the planet travels _____ in its orbit. The third law states
    that the _____ of the orbital period of the planet is
    proportional to the _____ of the average distance of the planet
    from the sun.

    It is important to remember that Kepler's laws are _____. In
    other words, Kepler determined them solely from _____.
    They were not _____ from a more fundamental law.

Tycho Brahe	five
Copernicus	geocentric
cube	heliocentric
derived	observations
Earth	perihelion
ellipses	slowest
empirical	square
equal	sun
fastest	three

Self-Test

Select the one best answer for each question. Check your answer with the Answer Key and review when necessary.

1. Eratosthenes based his measurement of the circumference of Earth on the observed position of the sun at the same time and on the same day in the cities of Alexandria and Syene. He observed that the sun at Syene was

   a. on the horizon and at the zenith at Alexandria.

   b. at the zenith and on the horizon at Alexandria.

   c. at the zenith and 7 degrees south of (below) the zenith at Alexandria.

   d. in the same position in the sky as it was in Alexandria.

2. The unit of distance Eratosthenes used to calculate the circumference of Earth was the

   a. stadium.

   b. degree.

   c. kilometer.

   d. number of degrees in a complete circle.

3. Aristotle concluded that Earth remained stationary because he was unable to observe parallax in the stars. Parallax is the

   a. motion of the celestial sphere around the stationary Earth.

   b. relationship between the velocity of a planet and its distance from Earth.

   c. change in the apparent position of an object due to a change in the location of the observer.

   d. change in the position of a star due to the season of the year during which the observation is made.

4. Ptolemy assumed that the planets moved through the perfect heavens in perfect motion. This perfect motion is called

   a. uniform circular motion.

   b. uniform accelerated motion.

   c. motion without parallax.

   d. retrograde motion.

5. In order to account for some of the observed peculiar motions of the planets, Ptolemy's model required that the planets move in

   a. small circles called deferents.

   b. straight lines at constant velocity.

   c. small circles called epicycles.

   d. the equant.

6. As a result of Earth's rotation, the stars appear to rise in the east and set in the west. If we observe the motion of the planets against the background stars, the planets usually appear to move from

   a. east to west.

   b. west to east.

   c. north to south.

   d. north to west.

7. Planets are observed to move, for a time, in retrograde motion. This motion describes the

    a. easterly drift of the planets.

    b. westerly drift of the stars.

    c. backward or easterly motion of the stars.

    d. backward or westerly motion of the planets.

8. In *De Revolutionibus*, Copernicus proposed that the sun was located at

    a. one of the foci of an ellipse.

    b. the center of the universe.

    c. directly above Earth.

    d. the equant.

9. The movements of the planets in the heliocentric system of Copernicus required

    a. elliptical orbits.

    b. parabolic orbits.

    c. straight-line motion.

    d. uniform circular motion.

10. Epicycles were used to describe the motions of the planets by

    a. Copernicus.

    b. Ptolemy.

    c. both Copernicus and Ptolemy.

    d. astronomers before the time of Ptolemy.

11. The Copernican model of the solar system eventually gained acceptance over the Ptolemaic model. The probable main reason for this acceptance was that the

    a. accuracy of the predictions were better than the Ptolemaic model.

    b. sun was placed at one of the foci of the planetary orbits.

    c. sun was placed at the center of the solar system.

    d. whole idea of the heliocentric system was pleasing to the eye and intellect.

12. The success of Tycho Brahe is attributed to his ability to
    a. make accurate observations with large and well-constructed instruments.
    b. construct theoretical models of physical phenomena.
    c. receive financial support from the king for his astronomical endeavors.
    d. make excellent observations with a telescope.

13. According to the Tychonic model of the solar system,
    a. Earth was the center of the universe and the planets orbited Earth.
    b. Earth was stationary, the sun and moon orbited Earth, and the planets orbited the sun.
    c. the sun was the center of the universe and the planets orbited the sun in elliptical paths.
    d. the sun was the center of the universe and the planets orbited the sun in circular paths.

14. According to Kepler's first law of planetary motion, the orbits of the planets are
    a. circles, with Earth located at the center.
    b. ellipses, with the sun located at one of the foci.
    c. ellipses, with Earth located at one of the foci.
    d. are circles, with the sun located at the center.

15. A planet moves fastest in its orbit when it is
    a. closest to Earth.
    b. farthest from the sun.
    c. at one of the foci.
    d. closest to the sun.

16. Galileo's most convincing observation that proved Venus orbited the sun was the discovery of
    a. moons in orbit around Jupiter.
    b. the highly reflective atmosphere of Venus.
    c. the phase cycle of Venus.
    d. evidence for the rotation of the planet.

17. Galileo observed and recorded

   a. sunspots, satellites around Jupiter, phases of Venus, and lunar craters and mountains.

   b. dark lines on Venus, phases of Jupiter, and lunar craters and mountains.

   c. satellites around Jupiter, phases of Venus, sunspots, and lunar craters and mountains.

   d. sunspots, mountains on Venus, lunar craters, and satellites around Jupiter.

18. A single conjecture or assertion that must be tested best describes a

   a. law.

   b. theory.

   c. hypothesis.

   d. model.

19. A model can best be described as a

   a. description of a natural phenomenon.

   b. representation of what actually exists in nature.

   c. correct physical representation of a conceptual idea.

   d. statement of fact that has been tested and shown to be correct.

20. A commonly accepted set of scientific ideas and assumptions is referred to as a scientific

   a. model.

   b. hypothesis.

   c. paradigm.

   d. theory.

Using What You've Learned

According to your own interests or your instructor's assignment, complete one or more of the following activities.

1. Follow the instructions below to construct an ellipse, the basic geometrical shape for the orbits of the planets, comets, and asteroids.

    a. Cut a piece of strong string about 10 inches in length and tie the ends together.

    b. Place a sheet of paper on a board and push two thumbtacks, about 2 inches apart, into the board. These locations are called the foci.

    c. Place the loop of string around the outside of the two thumbtacks.

    d. Use a pencil to stretch the piece of string into the shape of a triangle. The pencil is one vertex and the two thumbtacks are the other two vertices. The string forms the sides of the triangle.

    e. Move the pencil, keeping the string taut, and allow the string to guide the pencil to draw an ellipse. Keep the pencil in contact with the paper at all times and make sure the pencil and thumbtacks remain the vertices of a triangle and the string remains as the sides of the triangle. Continue in this manner until you have completed an ellipse. If part of the ellipse falls off the paper, shorten the string. Refer to Figure 4-10(a) on page 54 in the *Horizons* textbook.

    f. After drawing the ellipse, measure the major axis (the largest diameter through the two foci = 2a) and the distance between the two foci (2c).

    g. Calculate the eccentricity of the ellipse:

$$e = c/a$$

2. In the diagram of a planetary orbit that appears on page 54, locate and label:

    a. one possible position of the sun (focus)

    b. the major axis

    c. the position where the planet would have its slowest speed

    d. perihelion

    e. the position where the planet would have its greatest speed

    f. aphelion

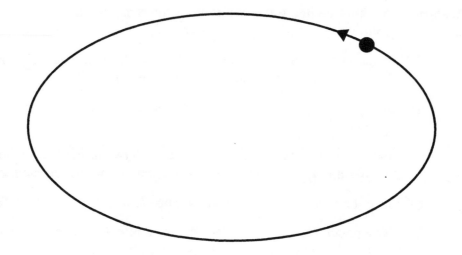

3. Develop a diagram that shows the occurrence of the phases of Venus. If Venus is currently visible, use a pair of binoculars to observe the planet and determine where on your diagram Venus must be in order for you to observe its current phase.

4. Where on Earth:

   a. do the stars rise in the west?

   b. is the south celestial pole at the south point on the horizon?

   c. is the sun at the north point on the horizon at midnight on the day of the summer solstice?

   d. is the zenith directly overhead?

   e. does the sun pass through the zenith on the day of the summer solstice?

   f. is the observer's latitude equal to the altitude of the north celestial pole (assume the observer is in the northern hemisphere)?

   g. does the sun rise once a year on the autumnal equinox?

   h. are all the stars above the horizon half of the time?

Lesson Review

Lesson 4

The Origin of Modern Astronomy

Please Note: Use this matrix to guide your study and achieve the learning objectives of this lesson. It will also help you to view the video, which defines and demonstrates important concepts and skills as they relate to everyday life.

Learning Objective	Textbook	Telecourse Student Guide	Background Notes
1. Describe how Eratosthenes measured the circumference of Earth.		Self-Test Questions: 1, 2.	Background Notes 4: Eratosthenes' Calculation of Earth's Circumference
2. Describe the geocentric models of the solar system developed by Aristotle and Ptolemy.	pp. 42–46	Matching Activities: 1, 2, 3, 5, 6, 7; Completion Activities: 1; Self-Test Questions: 3, 4, 5.	
3. Describe the motions of the naked-eye planets as observed from Earth and define the term retrograde.	pp. 44–45	Matching Activities: 4; Completion Activities: 1; Self-Test Questions: 6, 7.	
4. Explain the Copernican model and hypothesis.	pp. 46–50	Matching Activities: 8; Self-Test Questions: 8, 9.	
5. Compare and contrast the Ptolemaic (geocentric) model of the solar system with the Copernican (heliocentric) model.	pp. 43–50	Matching Activities: 8; Self-Test Questions: 10, 11.	
6. Identify the various instruments used by Tycho Brahe to make his observations and describe the various conclusions he made based on his observations.	pp. 50–52	Self-Test Questions: 12, 13.	

Learning Objective	Textbook	Telecourse Student Guide	Background Notes
7. Explain Kepler's three laws of planetary motion.	pp. 52–56	Matching Activities: 13; Completion Activities: 2; Self-Test Questions: 14, 15.	
8. Describe the telescopic observations of the solar system made by Galileo.	pp. 56–58	Self-Test Questions: 16, 17.	
9. Distinguish between paradigm, hypothesis, model, theory, and law.	pp. 14, 49, 55	Matching Activities: 9, 10, 11, 12, 14; Self-Test Questions: 18, 19, 20.	

5

Newton, Einstein, and Gravity

Assignments

For the most effective study of this lesson, we suggest that you complete the assignments in the sequence listed below.

Before Viewing the Video Program

- Read the Overview and Learning Objectives for this lesson. Use the Learning Objectives to guide your reading, viewing, and thinking.

- Read Chapter 4, "The Origin of Modern Astronomy," section 4-6, pages 60–66, in the *Horizons* textbook.

- Read Background Notes 5A, "The Law of Universal Gravitation," and 5B, "Special and General Relativity," following Lesson 6 in this guide.

- Read the Viewing Notes in this lesson.

View the "Newton, Einstein, and Gravity" Video Program

After Viewing the Video Program

- Briefly note your answers to questions listed at the end of the Viewing Notes.

- Review all reading assignments for this lesson, especially the Chapter 4 summary on page 66 in *Horizons*, Background Notes 5A and 5B for Unit I, and the Viewing Notes in this lesson.

- Respond to the challenges presented by each Critical Inquiry in the reading assignment and write brief answers to review questions 14 and 15 at the end of Chapter 4 in *Horizons* to be certain you understand the text material.

- Complete the Review Activities in this guide to reinforce your understanding of important terms and concepts. Check your answers with the Answer Key and review when necessary.

- Take the Self-Test in this guide to measure your achievement of the Objectives. Check your answers with the Answer Key and review when necessary.

- Complete the Using What You've Learned activities and any other activities and projects assigned by your instructor.

Overview

Johannes Kepler developed a set of mathematical rules that described how planets move around the sun. Galileo Galilei developed a set of mathematical rules that described how objects fall near the surface of Earth. Galileo's law of falling bodies was quite contrary to the ideas of motion held by Aristotle. Kepler's and Galileo's rules established the concept that mathematical relationships could be used to describe the behavior of the universe. However, these relationships were limited because they could describe only nearby events. They could not explain the causes of that motion. That next step in our understanding of the universe was made by Isaac Newton.

In his published work, the *Principia*, Newton put forth an explanation for the causes of motion. He introduced the ideas of force, mass, and acceleration. He redefined the nature of objects in motion and at rest. Newton spent nearly two decades developing his laws of motion, making fundamental discoveries in optics and mathematics and deriving a law that describes one of the most, if not the most important force in the universe: gravity. It is through understanding and applying gravity that scientists have been able to calculate the return of comets, predict the presence of unseen planets, land humans on the moon, and launch spacecraft that escape from the solar system.

Along with our ability to describe the universe with mathematical relationships, we also developed the technology to observe the universe in greater detail. It is the result of these observations that brought the laws developed by Isaac Newton into question. For instance, Newton's law of universal gravitation could not satisfactorily explain some phenomena, and modern science found inconsistencies between Newton's laws of motion and the behavior of small atomic particles traveling at high speeds. It was the general and special theories of relativity of Albert Einstein that explained the new observations and resolved the inconsistencies. Because Einstein's theories required few fundamental postulates, the theories are aesthetically pleasing to scientists. However simple these postulates seem to be, they are very powerful and provide us with insight into areas beyond the realm of our personal experiences and intuition. The logical, observational, and experimental evidence in support of these theories has elevated them to the status of laws.

This entire lesson is devoted to developing an understanding of the classical (Newton) and modern (Einstein) view of the universe. Newton's laws of motion and Einstein's theories of relativity describe the nature of the universe, can be expressed mathematically, and are supported by experimental evidence. With these laws as a foundation, we can now explore everything from the nature of planets and their orbits about the sun to the nature of stars and their evolution. We can even begin to develop a description and understanding of the origin, evolution, and destiny of the universe.

Learning Objectives

When you have completed all assignments for this lesson, you should be able to:

1. Describe Newton's three laws of motion. *HORIZONS* TEXTBOOK PAGES 61–62.

2. State Newton's law of universal gravitation and describe how it explains orbital motion. *HORIZONS* TEXTBOOK PAGES 61–65; BACKGROUND NOTES 5A.

3. Identify the various astronomical observations and calculations performed in the eighteenth and nineteenth centuries that initially supported the validity of Newton's law of universal gravitation. BACKGROUND NOTES 5A; VIDEO PROGRAM.

4. State the basic postulates of Einstein's special theory of relativity. BACKGROUND NOTES 5B; VIDEO PROGRAM.

5. Explain and describe the following consequences of special relativity: time dilation and length contraction. BACKGROUND NOTES 5B; VIDEO PROGRAM.

6. Explain the relationship $E = mc^2$. BACKGROUND NOTES 5B; VIDEO PROGRAM.

7. State the basic postulates of Einstein's general theory of relativity. BACKGROUND NOTES 5B; VIDEO PROGRAM.

8. Explain what is meant by the curvature of space-time and describe the following consequences of general relativity: precession of orbits (Mercury), bending of starlight, time dilation and gravitational red shift, and gravitational radiation (gravity waves). BACKGROUND NOTES 5B; VIDEO PROGRAM.

Viewing Notes

"Newton, Einstein, and Gravity" is about the quest of science to understand the nature of gravity and motion. The first part of the video program traces how early scientists, including Aristotle, Galileo, and Kepler, tried to explain gravity and motion.

Most of the video program focuses on the contributions of two scientists: Isaac Newton and Albert Einstein. Newton—in his universal law of gravity and three laws of motion—was the first person to explain successfully such phenomena as elliptical orbits and the motions of all objects. Film clips of astronauts in an orbiting space shuttle clearly illustrate Newton's three laws of motion and their universality. Computer graphics help to explain Newton's other major contribution: the law of universal gravitation and how it describes orbits.

Although Newton's laws worked for the universe as it was known in his time, they did not explain phenomena that scientists were able to observe around the beginning of the twentieth century, such as atoms and electrons moving at near the speed of light. It was Einstein who developed theories to explain what Newton's laws could not explain.

In the video program, several physicists and astronomers explain Einstein's special and general theories of relativity. Computer graphics illustrate key aspects of each theory, including the universality of the speed of light, time dilation, the twin paradox, equivalence of acceleration and gravity, and the curvature of space-time.

The video program concludes with two striking "proofs" of Einstein's theories: how the general theory of relativity explains the precession of Mercury's orbit and how the sun's gravity bends light from stars behind it.

As you watch the video program, consider the following questions:

1. What is the difference between a theory and a law?

2. What are Newton's three laws of motion? What is Newton's law of universal gravitation?

3. What is the principle of relativity? What did Einstein say about the speed of light as measured by different observers?

4. What is the twin paradox?

5. How can muons reach the surface of Earth?

6. What do we mean when we say that gravity and acceleration are equivalent?

7. What do we mean by space-time? What do we mean when we say that mass curves space-time around it?

8. How was a solar eclipse used to observe the bending of starlight around the sun?

Review Activities

Matching

Match the terms listed below with the definitions that follow. Check your answers with the Answer Key and review any terms you missed.

____ 1. mass

____ 2. weight

____ 3. inverse square law

____ 4. *Principia*

____ 5. Newton's first law of motion

____ 6. Newton's second law of motion

____ 7. Newton's third law of motion

____ 8. law of universal gravitation

____ 9. space-time

____ 10. time dilation

____ 11. $E = mc^2$

____ 12. length contraction

____ 13. gravitational red shift

____ 14. gravity waves

____ 15. curvature of space-time

____ 16. principle of equivalence

a. the slowing of time as measured by one observer in motion relative to another

b. a rule that the strength of an effect (such as gravity) decreases in proportion as the distance squared increases

c. ripples in space-time that may travel outward from sources of gravity

d. a measure of the amount of matter in an object

e. the shortening of length in the direction of motion as measured by an observer considered to be at rest

f. the warping of space in the presence of mass

g. when one body exerts a force on another, the second body exerts an equal but opposite force on the first body

h. a body continues at rest or in uniform motion in a straight line unless acted upon by some force

i. the four dimensions in which all things (matter and energy) exist; one dimension for time and the other three for possible directions (motions) in space

j. the lengthening of the wavelength of a photon (decrease in frequency) due to its escape from a gravitational field

k. the force of gravity acting on an object on or near a planet or moon; it varies from place to place and planet to planet

l. relationship between mass and energy

m. the force between two objects is determined by their masses and the inverse square of the distance between them

n. publication by Isaac Newton containing his work on laws of motion and universal gravitation

o. a body's change of motion is proportional to the force acting on it and is in the direction of the force

p. basic postulate of general relativity describing the relationship between a gravitational force and a constant acceleration

Completion

Fill each blank with the most appropriate term from the list for that paragraph. A term may be used once, more than once, or not at all. Check your answers with the Answer Key and review when necessary. If a question requires two or more answers in succession, they may be in any order, unless the question indicates otherwise.

1. Isaac Newton developed, among other things, three laws to explain motion and one law to describe gravity. His first law describes the nature of motion of an object. It states that if an object is at rest it will remain at _____. If the object is in motion, it will _____ in motion at a _____ speed and in a constant _____ until it is acted upon by some _____.
According to Newton's second law, the _____ in a body's motion is _____ to the force acting on the body. Furthermore, that _____ in the motion is in the _____ direction as the force acting on the body. If a force is applied to an object, the object exerts a _____, _____ in size to the original _____ but _____ in direction, on the first body. This _____ law is known as Newton's third law of motion.
Newton explained the nature of gravity in his law of _____ gravitation. The force of gravity is _____ to the _____ of the objects and _____ proportional to the _____ of the distance between them.

action-reaction	masses
change	opposite
constant	proportional
continue	rest
conversely	same
direction	special
equal	square
force	universal
inversely	variable

2. According to Einstein's special theory of relativity, the laws of physics are the _____ for all observers, and the _____ of _____ is also the same for all observers regardless of their relative motions. These two fundamental postulates require us to reevaluate the nature of _____ and _____. The effects of special relativity are important when objects travel at speeds near that of _____. The phenomenon of time _____ results in the _____ of clocks in a _____ spacecraft when compared with similar clocks that are considered to be at _____. An example of this phenomenon is found in the motion of the subatomic particle called the _____. Its lifetime is so _____, it should not be able to reach the surface of Earth when produced in the upper atmosphere. However, because of _____, the clock of the _____ runs _____ and the particle lives long enough to reach the ground. Another important result of special relativity demonstrates the relationship between mass and _____. In the famous equation $E = mc^2$, the "E" stands for _____, the "m" stands for _____, and the "c" stands for the speed of _____. Once this equation was understood, scientists could explain why the _____ shine for as long as they do.

brief	rest
dilation	same
$E = mc^2$	slower
energy	slowing
faster	space
light	speed
long	stars
mass	time
moving	time contraction
muon	time dilation

3. Einstein's general theory of relativity is based on the postulate that equates _____ and _____. This statement is called the _____. It states that it is _____ to distinguish between a constant _____ and being at rest in a _____ field. General relativity is our best way to describe gravity. It describes how gravity produces a _____ of space-time and that space-time consists of _____ dimensions. General relativity also explains why the orbit of Mercury _____ around the sun, how gravity can _____ starlight, and how time _____ in a gravitational field. The effect gravity has on time is known as a gravitational _____. Furthermore, general relativity makes a prediction about the existence of _____ waves. These waves are ripples in _____ and travel outward, at the speed of _____, from the explosion of a star or from two stars in orbit about each other. Although these waves have yet to be detected directly, their existence is indirectly verified by the observation of two _____ stars in orbit about each other. The orbit slowly _____ by the amount predicted by general relativity.

acceleration	light
bend	neutron
curvature	precesses
decaying	principle of equivalence
decays	red shift
four	six
gravitational	slows
gravity	space-time
impossible	two
increases	

Self-Test

Select the one best answer for each question. Check your answers with the Answer Key and review when necessary.

1. If an object is moving and it remains undisturbed by outside forces, it will
   a. stop.
   b. continue to move in a circular path.
   c. continue to move in a straight line with a uniform velocity.
   d. stop and begin to move in the opposite direction.

2. If you double the force acting on an object, assuming the mass remains constant, the acceleration of that object would be
   a. doubled.
   b. halved.
   c. four times greater.
   d. one-fourth as great.

3. According to Newton's law of universal gravitation, if the distance between two objects increases, the force of gravity between them will
   a. increase.
   b. decrease by a factor equal to the square of the distance between the objects.
   c. increase by a factor equal to the square of the distance between the objects.
   d. remain unchanged.

4. A satellite launched into orbit around Earth, far above the effects of Earth's atmosphere, will

   a. eventually escape Earth's gravity and orbit the moon.

   b. eventually escape Earth's gravity and go into orbit around the sun.

   c. continue to orbit Earth.

   d. remain fixed in space as Earth rotates beneath the satellite.

5. The discovery of Neptune was based on the observed gravitational perturbations of the planet

   a. Mars.

   b. Jupiter.

   c. Saturn.

   d. Uranus.

6. One of the earliest uses of Newton's law of universal gravitation was the calculation of the orbit of a comet by

   a. Edmund Halley.

   b. Isaac Newton.

   c. John C. Adams.

   d. William Herschel.

7. The speed of light

   a. varies with distance from the observer.

   b. remains constant for all observers.

   c. decreases with time.

   d. increases when two objects are observed to be approaching each other.

8. Although observers may move at uniform speeds relative to each other, they will always find that

   a. time is absolute.

   b. mass cannot transform into energy.

   c. the laws of physics depend on the relative motions of observers.

   d. the laws of physics are the same for all observers regardless of their relative uniform motion.

9. An example of time dilation can be observed by the fact that

    a. muons survive the journey from the upper atmosphere to the surface of Earth.

    b. mass can be transformed into energy.

    c. Halley's comet orbits the sun once every 76 years.

    d. the Michelson-Morely experiment did not detect the ether.

10. According to the special theory of relativity, *moving* clocks

    a. run more slowly at all times.

    b. run more slowly when the observer is moving at extremely high velocities.

    c. run more slowly when compared to other clocks that are considered to be at rest.

    d. never run more slowly. The effect is an illusion. Time runs at the same rate for everyone.

11. The "*E*" in the relationship $E = mc^2$ refers to the energy

    a. produced if all the mass "*m*" is converted to energy.

    b. associated with the object's motion.

    c. absorbed in an atomic explosion.

    d. produced in a nuclear power plant.

12. One of the most immediate applications of the equation $E = mc^2$ to astronomical phenomena was to explain the source of energy in

    a. orbiting planets.

    b. the interior of stars.

    c. muons and electrons.

    d. black holes and neutron stars.

13. "It is impossible to distinguish between a uniform acceleration and being at rest in a gravitational field" is a statement of the

    a. time dilation effect.

    b. principle of equivalence.

    c. length contraction phenomenon.

    d. curvature of space-time.

14. Space-time is said to consist of four dimensions. These four are
    a. two time dimensions and two space dimensions.

    b. three time dimensions and one space dimension.

    c. three space dimensions and one time dimension.

    d. four space dimensions.

15. In the absence of mass, space-time is said to be flat. In the presence of mass, space-time is said to be
    a. smooth.

    b. length contracted.

    c. confused with accelerations.

    d. curved.

16. According to Newton's law of universal gravitation, a beam of light is considered to have no mass and therefore would be unaffected by gravity. General relativity, however, predicts that light will be affected by gravity. This prediction is demonstrated by the fact that light
    a. travels only in straight lines.

    b. is observed to have a constant speed.

    c. creates length contraction effects.

    d. bends around massive objects.

17. Hypothetical ripples in space-time produced by supernova events or orbiting massive stars, which propagate outward at the speed of light, are referred to as
    a. gravity waves.

    b. neutron stars.

    c. black holes.

    d. gravitational red shifts.

Using What You've Learned

Provide short answers to the following questions about special and general relativity.

1. Why was the Michelson-Morely experiment considered a failure?

2. How did G. F. Fitzgerald explain the failure of the Michelson-Morely experiment?

3. What is the fundamental postulate concerning the speed of light in Einstein's special theory of relativity?

4. What is meant by space-time?

5. State the principle of equivalence in your own words.

6. Take a ride on a roller coaster, such as the "Star Tours" ride at Disneyland or Disney World, or any "multimedia" ride that simulates motion with visual aids. List the various times during the ride when accelerations simulate gravity and when gravity simulates accelerations.

7. What are the important differences between special and general relativity?

8. Pay close attention to your activities for a period of several hours. Make a list of ten significant activities you did during this time. Explain how Newton's three laws of motion related to your activities.

9. Watch several cartoon programs such as the *Road Runner* and *Tom and Jerry*. Identify the various scenes that violate Newton's laws of motion. Explain what would happen to the cartoon characters if the laws were allowed to occur correctly. (Ignore injuries and possible death to the cartoon characters.)

Lesson Review

Lesson 5

The Origin of Modern Astronomy

Please Note: Use this matrix to guide your study and achieve the learning objectives of this lesson. It will also help you to view the video, which defines and demonstrates important concepts and skills as they relate to everyday life.

Learning Objective	Textbook	Telecourse Student Guide	Background Notes
1. Describe Newton's three laws of motion.	pp. 61–62	Matching Activities: 1, 2, 3, 4, 5, 6, 7; Completion Activities: 1; Self-Test Questions: 1, 2.	
2. State Newton's law of universal gravitation and describe how it explains orbital motion.	pp. 61–65	Matching Activities: 1, 2, 3, 8; Completion Activities: 1; Self-Test Questions: 3, 4.	Background Notes 5A: The Law of Universal Gravitation
3. Identify the various astronomical observations and calculations performed in the eighteenth and nineteenth centuries that initially supported the validity of Newton's law of universal gravitation.		Self-Test Questions: 5, 6.	Background Notes 5A: The Law of Universal Gravitation
4. State the basic postulates of Einstein's special theory of relativity.		Matching Activities: 9; Completion Activities: 2 Self-Test Questions: 7, 8.	Background Notes 5B: Special and General Relativity
5. Explain and describe the following consequences of special relativity: time dilation and length contraction.		Matching Activities: 10, 12; Completion Activities: 2; Self-Test Questions: 9, 10.	Background Notes 5B: Special and General Relativity
6. Explain the relationship $E = mc^2$.		Matching Activities: 11; Completion Activities: 2; Self-Test Questions: 11, 12.	Background Notes 5B: Special and General Relativity

Learning Objective	Textbook	Telecourse Student Guide	Background Notes
7. State the basic postulates of Einstein's general theory of relativity.		Matching Activities: 16; Completion Activities: 3; Self-Test Questions: 13, 14.	Background Notes 5B: Special and General Relativity
8. Explain what is meant by the curvature of space-time and describe the following consequences of general relativity: precession of orbits (Mercury), bending of starlight, time dilation and gravitational red shift, and gravitational radiation (gravity waves).		Matching Activities: 13, 14, 15; Completion Activities: 3; Self-Test Questions: 15, 16, 17.	Background Notes 5B: Special and General Relativity

The Tools of Astronomy

Assignments

For the most effective study of this lesson, we suggest that you complete the assignments in the sequence listed below.

Before Viewing the Video Program

- Read the Overview and Learning Objectives for this lesson. Use the Learning Objectives to guide your reading, viewing, and thinking.

- Read Chapter 5, "Astronomical Tools," pages 68–91, in the *Horizons* textbook.

- Read the Viewing Notes in this lesson.

View "The Tools of Astronomy" Video Program

After Viewing the Video Program

- Briefly note your answers to questions listed at the end of the Viewing Notes.

- Review all reading assignments for this lesson, especially the Chapter 5 summary on pages 90 in *Horizons*, and the Viewing Notes in this lesson.

- Respond to the challenges presented by each Critical Inquiry in the reading assignment and write brief answers to all the review questions at the end of Chapter 5 in *Horizons* to be certain you understand the text material.

- Complete the Review Activities in this guide to reinforce your understanding of important terms and concepts. Check your answers with the Answer Key and review when necessary.

- Take the Self-Test in this guide to measure your achievement of the Objectives. Check your answers with the Answer Key and review when necessary.

- Complete any other activities and projects assigned by your instructor.

Overview

The course thus far has been almost entirely about the history of astronomy—how humanity's quest to understand the universe was transformed from an almost mystical preoccupation in ancient times to the first modern science during the Renaissance. Even when discussing natural phenomena such as eclipses and seasons, the course has concentrated on naked-eye phenomena that were well known to our distant ancestors and were an integral part of the reason they chose to study the sky.

It should be clear that the turning point in astronomy was Galileo's use of the telescope in the early seventeenth century. Although his instrument was crude compared to even a toy telescope today, it still revealed the cosmos with about ten times more clarity than the unaided human eye can discern. Consequently, almost everything he looked at was a major, historical discovery, and often a contradiction of some cherished ancient belief. Although professional telescopes today are vastly larger and more powerful than Galileo's, this capability has grown slowly, in small increments, over centuries. Thus, Galileo's humble spyglass remained the single greatest technological leap forward in astronomical resolving power, until the successful repair of the Hubble Space Telescope accomplished a similarly great leap late in 1993.

Today, the telescope is the single most important research device available to astronomers who, unlike other scientists, can only look at, but not touch, the objects they study. It is actually the light from distant stars which astronomers touch; it is the light which has physically traveled through space to Earth, which they have in their possession, and which they concentrate and analyze with their instruments. Observational astronomy is a quest for light. This lesson shows that every improvement in astronomical technology has been the result of an attempt either to gather more light, to gather it over a wider range of wavelengths, or to focus it in a way which reveals greater detail. And, as you will see throughout the remainder of the course, every improvement in technology has led to a corresponding improvement in our understanding of the universe.

Learning Objectives

When you have completed all assignments for this lesson, you should be able to:

1. Explain the characteristics of the electromagnetic spectrum and their effects on astronomical equipment and observations. *HORIZONS* TEXTBOOK PAGES 69–71; VIDEO PROGRAM.

2. Describe the operation of optical telescopes, including the relative advantages and disadvantages of refracting versus reflecting designs, different focus locations, and mountings. *HORIZONS* TEXTBOOK PAGES 71–81.

3. Describe the factors that determine the light-gathering, magnifying, and resolving powers of a telescope, including the effects of Earth's atmosphere. *HORIZONS* TEXTBOOK PAGES 73–77; VIDEO PROGRAM.

4. Describe the instrumentation used to accomplish photography, photometry, and spectroscopy with a telescope. *HORIZONS* TEXTBOOK PAGES 81–83; VIDEO PROGRAM.

5. Describe the operation, limitations, and advantages of radio telescopes, including the principle of interferometry. *HORIZONS* TEXTBOOK PAGES 84–87; VIDEO PROGRAM.

6. Describe space-based telescopes designed for both visible and nonvisible astronomy. *HORIZONS* TEXTBOOK PAGES 87–90; VIDEO PROGRAM.

Viewing Notes

This video program, "The Tools of Astronomy," is quite different from most of the others in this telecourse. Normally, the programs use animation and the personal descriptions of astronomers' own research to help the student visualize the major concepts and facts that are also described in the text. This program is more of a supplement to the text; it uses the "you are there" power of television to enable you to visit some of the greatest observatories of the past and present and to see the environment in which astronomers carry out their work.

The first segment begins, academically enough, with an overview of the electromagnetic spectrum, from gamma rays to radio waves, and the types of information provided by each region. But immediately, more practical concerns are described. Visible light is only a small part of that spectrum, and many natural phenomena in space emit most of their radiation at nonvisible wavelengths. Not only does this require new technology to detect, but we are faced with the problem that Earth's own atmosphere is opaque to most of these wavelengths. Even for visible light, the turbulence of the atmosphere distorts the images produced by ordinary, optical telescopes located here on the ground.

For that reason, as the second video segment explains, astronomers a century ago began to move to mountain tops to avoid some of that distortion. The first such telescope was the 36-inch refractor at Lick Observatory. But the lens of a refractor cannot be made much larger than that, so astronomers turned to reflectors and soon constructed the 100-inch Hooker reflector on Mount Wilson, and then the 200-inch Hale reflector on Palomar Mountain.

These telescopes are all still in use. But to see the latest in telescope designs, viewers are given a tour of the 4-meter Mayall reflector at Kitt Peak National Observatory, and the awesome 10-meter Keck reflector on Mauna Kea in Hawaii. This latter instrument almost doubled the diameter of the previously largest optical telescope by making a huge mirror of many smaller segments, and was constructed at what may be the best observing site on the planet as well. (Since the video was produced, a second identical Keck telescope has been constructed right next to the first.)

The segment concludes by showing that ever larger telescopes are not the only way to improve observations. The replacement of chemical photography by digital electronic charge-coupled devices (CCDs) has effectively turned smaller telescopes into the functional equivalent of much larger ones by greatly speeding up the process of gathering and recording images from light. And adaptive optics, with its computer-controlled "rubber" mirrors, promises to compensate for much of the distortion of Earth's atmosphere, which has so troubled astronomers from Galileo on.

The third segment begins to wander elsewhere along the electromagnetic spectrum by describing radio telescopes. Radio waves are the only other wavelength region, besides visible light, to which the atmosphere is transparent so that observations can be carried out on the ground. But the initially poor resolution, or resolving power, of radio images prompted the development of interferometry so that radio arrays today can actually reveal finer detail than optical telescopes.

The final segment follows telescopes into space to study the higher energy wavelengths, such as gamma rays and X-rays, which cannot reach the ground. This segment features a look at the Extreme Ultraviolet Explorer satellite and its "grazing incidence" telescopes. Since extreme ultraviolet light cannot be focused by ordinary lenses or mirrors, these telescopes use what amounts to converging hollow cylinders, with reflective coatings along the inside walls, to funnel the light as it strikes the walls at shallow angles.

As you watch the video program, consider the following question:

With regard to the three major goals of (a) gathering more light, over (b) a wider range of wavelengths, with (c) finer resolution, which goal was the primary accomplishment of each of these innovations?

Switching from refraction to reflection for optical telescopes

Moving optical telescopes to mountain tops

Making optical mirrors of many multiple segments

Developing adaptive optics

Using charge-coupled devices

Inventing radio telescopes

Inventing radio interferometry

Placing nonoptical telescopes in space

Developing grazing incidence reflectors

Placing optical telescopes, such as the Hubble Space Telescope, in space

Review Activities

Matching

Match the terms listed below with the definitions that follow. Check your answers with the Answer Key and review any terms you missed.

____ 1. electromagnetic radiation

____ 2. wavelength

____ 3. photon

____ 4. atmospheric window

____ 5. refracting telescope

____ 6. focal length

____ 7. objective

____ 8. eyepiece

____ 9. chromatic aberration

____ 10. reflecting telescope

____ 11. prime focus

____ 12. Cassegrain focus

____ 13. Newtonian focus

____ 14. active optics

____ 15. equatorial mounting

____ 16. alt-azimuth mounting

____ 17. light-gathering power

____ 18. resolving power

____ 19. seeing

____ 20. magnifying power

____ 21. charge-coupled device (CCD)

____ 22. spectrograph

____ 23. grating

____ 24. radio interferometer

a. mounting in which one axis points to a celestial pole

b. mounting in which one axis points vertically upward

c. a quantum of electromagnetic energy

d. fields that transfer energy, such as light waves, radio waves, and gamma rays

e. inversely related to the energy of a photon

f. wavelength region in which a particular photon can reach the ground from space

g. computerized control of the shape of a mirror

h. the point at which the objective mirror forms an image

i. a focus location at the upper side of a telescope

j. a focus location directly beneath the objective mirror

k. the lens or mirror that first produces a focused image, as opposed to modifying, enlarging, or redirecting the image

l. a telescope that has a lens for an objective

m. a telescope that focuses light with a mirror

n. a lens for visually enlarging a focused image

o. a color distortion peculiar to simple lenses but not to mirrors

p. the distance between an objective and its focused image of a very distant object

q. combining of signals from two or more radio telescopes to increase resolving power

r. an electronic device for recording faint images

s. an alternative to a prism as a way of dispersing light into its component colors

t. an alternative to a prism as a way of dispersing light into its component colors

u. a term for the quality of atmospheric conditions on a given night

v. the ability of a telescope to make starlight brighter

w. the ability of a telescope to make an image larger

x. the ability of a telescope to reveal fine detail

Completion

Fill each blank with the most appropriate term from the list for that paragraph. A term may be used once, more than once, or not at all. Check your answers with the Answer Key and review when necessary.

1. The energy of a photon is related to its wavelength. A photon with a short wavelength has _____ energy than a photon with a longer wavelength. For example, in comparison to an ultraviolet photon, an infrared photon has _____ energy. One problem faced by astronomers is that only a small portion of electromagnetic radiation reaches Earth. Earth's uppermost atmosphere absorbs _____ rays, _____, and some _____ waves. About 30 kilometers above Earth's surface, a layer of ozone absorbs _____ radiation. In the lower atmosphere, water vapor absorbs _____ radiation.

gamma	radio
greater	ultraviolet
infrared	X rays
lesser	

2. In a refracting telescope, the objective is a _____. The lens closest to the object being viewed is called the _____ lens. The light-gathering power and resolving power of a telescope depend on the size of the objective lens; with a larger diameter lens, those powers are _____. In a telescope, when observing visually, the image is magnified by the _____. One of the main problems of refracting telescopes is the inability to focus all colors of light simultaneously, which is known as _____ aberration.

chromatic	lesser
eyepiece	mirror
greater	objective
lens	

3. In a reflecting telescope, the objective is a _____. The image is formed at the prime focus at the _____ end of the tube and then reflected by the _____ to a place where it can be viewed more easily. One type of reflecting telescope, a Schmidt camera, can

correct distortions in an image with a thin _____. One of the principal advantages of reflecting telescopes is that light does not pass through a significant thickness of _____.

glass	mirror
lens	secondary mirror
lower	upper

4. In a radio telescope, the function performed by the mirror in a reflecting telescope is often performed by a _____ reflector. In comparison to other types of telescopes, the resolving power of a radio telescope is _____. Astronomers overcome this problem by linking radio telescopes together to form a radio _____. In such an arrangement, the resolving power increases to that of a single telescope with a diameter _____ to the distance between the dishes.

antenna	equal
better	interferometer
dish	poorer

Self-Test

Select the one best answer for each question. Check your answers with the Answer Key and review when necessary.

1. Of the types of electromagnetic radiation listed below, the one with the shortest wavelength is

   a. X ray.

   b. radio wave.

   c. visible light.

   d. ultraviolet light.

2. In addition to visible light, Earth's atmosphere is transparent to

   a. X rays.

   b. radio waves.

   c. gamma rays.

   d. infrared radiation.

3. The 200-inch Hale Telescope on Palomar Mountain is
   a. the largest refracting telescope.

   b. one of the largest refracting telescopes.

   c. the largest reflecting telescope.

   d. one of the largest reflecting telescopes.

4. The newest, largest ground-based optical telescopes are supported by
   a. equatorial mountings.

   b. alt-azimuth mountings.

   c. stationary mountings with moving coelostat mirrors.

   d. none of the above.

5. For an optical telescope located on Earth under normal seeing conditions, the main value of making ever larger telescopes is improved
   a. magnifying power.

   b. resolving power.

   c. light-gathering power.

   d. air quality.

6. If the eyepiece of a telescope is replaced with an eyepiece with a focal length twice as long, the magnifying power is
   a. four times greater.

   b. two times greater.

   c. four times less.

   d. two times less.

7. A device that analyzes light by spreading it out according to wavelength is called
   a. an interferometer.

   b. a camera.

   c. a photometer.

   d. a spectrograph.

8. A device that disperses light into its component colors using thousands of microscopic parallel grooves is called a

   a. prism.

   b. grating.

   c. CCD.

   d. lens.

9. The advantages of a radio telescope over a large reflecting optical telescope include all of the following **EXCEPT** the ability to

   a. reveal finer detail.

   b. detect cool hydrogen.

   c. see through interstellar dust.

   d. see some of the most distant objects in the universe.

10. The main reason for creating an interferometer with two separate radio dishes is to

    a. monitor two different radio wavelengths simultaneously.

    b. increase the intensity of the detected signal.

    c. compare objects in two different directions simultaneously.

    d. increase the resolution in the data.

11. A high-altitude airplane is used to make observations

    a. at gamma-ray wavelengths.

    b. at X-ray wavelengths.

    c. in the ultraviolet.

    d. in the infrared.

12. The Hubble Space Telescope is

    a. a gamma-ray telescope.

    b. an X-ray telescope.

    c. an optical telescope.

    d. a radio telescope.

Lesson Review

Lesson 6

The Tools of Astronomy

Please Note: Use this matrix to guide your study and achieve the learning objectives of this lesson. It will also help you to view the video, which defines and demonstrates important concepts and skills as they relate to everyday life.

Learning Objective	Textbook	Telecourse Student Guide	Background Notes
1. Explain the characteristics of the electromagnetic spectrum and their effects on astronomical equipment and observations.	pp. 69–71	Matching Activities: 1, 2, 3, 4; Completion Activities: 1; Self-Test Questions: 1, 2.	
2. Describe the operation of optical telescopes, including the relative advantages and disadvantages of refracting versus reflecting designs, different focus locations, and mountings.	pp. 71–81	Matching Activities: 5, 6, 7, 8, 9, 10, 11, 12, 13, 14, 15, 16; Completion Activities: 2, 3; Self-Test Questions: 3, 4.	
3. Describe the factors that determine the light-gathering, magnifying, and resolving powers of a telescope, including the effects of Earth's atmosphere.	pp. 73–77	Matching Activities: 17, 18, 19, 20; Self-Test Questions: 5, 6.	
4. Describe the instrumentation used to accomplish photography, photometry, and spectroscopy with a telescope.	pp. 81–83	Matching Activities: 21, 22, 23; Self-Test Questions: 7, 8.	
5. Describe the operation, limitations, and advantages of radio telescopes, including the principle of interferometry.	pp. 84–87	Matching Activities: 24; Completion Activities: 4; Self-Test Questions: 9, 10.	

Learning Objective	Textbook	Telecourse Student Guide	Background Notes
6. Describe space-based telescopes designed for both visible and nonvisible astronomy.	pp. 87–90	Self-Test Questions: 11, 12.	

Unit I

Background Notes

Scientific Method

One of the great challenges in the science of astronomy is to study objects and phenomena at very great distances from us. Unlike biology or zoology, astronomers usually cannot bring what they study into the laboratory for closer analysis. Therefore, in their observations and investigations, it is critical for astronomers to follow a consistent and logical process in developing explanations for the phenomena being studied. Adherence to this process is vital in all sciences because it allows scientists and the general public to have confidence that the knowledge gained describes the physical universe as it really is, to the best of human knowledge at a given point of time. This process not only gives us confidence in our understanding of nature but also allows us to discover laws that govern the behavior of nature. This logical and consistent process is called the scientific method.

In the scientific method, investigation or study of a particular object or phenomenon begins with observation. Scientists record information about the subject under study. For astronomers, this information may be in the form of numerical data or photographic images made from visible, X-ray, or infrared light. Because of technology, astronomers don't have to see with their eyes everything they measure.

After completing their observations, scientists organize their findings and use them to develop an explanation or proposal to account for the phenomenon being studied. This initial explanation is called a **hypothesis**. A hypothesis must be worded in such a way that it permits other scientists to test the explanation. Scientists conduct experiments and acquire additional observations to see if a hypothesis is reasonable. A hypothesis should also be such that it can be used to predict new observations and the results of new or different experiments. Furthermore, a hypothesis cannot violate or contradict already known and accepted facts. It must be consistent with what has already been validated. Even though a hypothesis may break new ground or introduce a new way of looking at things, it cannot disagree with what is already verified. A hypothesis must expand and enhance knowledge.

If a hypothesis is tested extensively and found to be true, it may be elevated to the status of a **model** or **theory**. A model is a picture of a natural phenomenon. For example, the model of the atom as an object consisting of a nucleus with electrons in orbit about it is a picture of the smallest structure in the universe. Although, at this time, we are not certain this "picture" of the atom is correct, the model enables us to understand and predict the atom's behavior. In contrast to a model, a theory is a set of rules that describes a particular phenomenon. Also, a theory can be applied to a large number of other situations and predict behavior. A theory is not limited to one event or object. To be accepted, a theory has to be tested, evaluated, expanded, and made more general.

Once a model or theory has been subjected to extensive testing and experimentation, scientists can begin to feel that they know the "truth" about an object or phenomenon. This truth is not absolute. It is probably better to say that the scientist has great confidence in what is known, or that the information in the model or theory represents the best of our present knowledge. The scientist must also be aware that the scientific method will be used to investigate further and that this "truth" may change and our best knowledge will change. This ongoing search for "truth" is the process of science and how we learn about the universe.

B A C K G R O U N D   N O T E S   2 A

The Roots of Astronomy

Astronomy has its origin in that most noble of all human traits, curiosity. Just as modern children ask their parents what the stars are and why the moon changes, so did ancient humans ask themselves these questions. The answers, often couched in mythical or religious terms, reveal a great reverence for the order of the heavens.

The study of the astronomy of ancient peoples, **archaeoastronomy,** came to public attention in 1965 when Gerald Hawkins published a book called *Stonehenge Decoded.* He reported that Stonehenge, the prehistoric ring of stones, was a sophisticated astronomical observatory.

Stonehenge, standing on Salisbury Plain in southern England, was built in stages from about 2800 B.C. to about 1075 B.C., a period extending from the late Stone Age into the Bronze Age. Though the public is most familiar with the massive stones of Stonehenge, those were added late in its history. In its first stages, Stonehenge consisted of a circular ditch slightly larger in diameter than a football field with a concentric bank just inside the ditch and a long avenue leading away toward the northeast. A massive stone, the Heelstone, stood then, as it does now, outside the ditch in the opening of the avenue.

As early as A.D. 1740, the English scholar W. Stukely suggested that the avenue pointed toward the rising sun at the summer solstice, but few accepted the idea. More recently, astronomers have recognized significant astronomical alignments at Stonehenge. Seen from the center of the monument, the summer solstice sun rises behind the Heelstone. Other sight lines point toward the most northerly and most southerly risings of the moon.

Michael A. Seeds, *Foundations of Astronomy* [Belmont, Calif.: Brooks/Cole, 2003], reprinted by permission of the publisher.

The significance of these alignments has been debated. Hawkins claims that the Stone Age people who built Stonehenge were using it as a device to predict lunar eclipses. Others have been more conservative, but the truth may never be known. The builders of Stonehenge had no written language and left no records of their intentions. Nevertheless, the presence of solar and lunar alignments at Stonehenge and at many other Stone Age monuments dotting England and Europe shows that so-called primitive peoples were paying detailed attention to the sky. The roots of astronomy lie not in sophisticated science and logic but in human curiosity and wonder.

The early inhabitants of North America were also interested in astronomy. The Big Horn Medicine Wheel (*medicine* is used here to mean magical power), on a 3000-meter shoulder of the Big Horn Mountains of Wyoming, was used by the Plains Indians about A.D. 1500–1750. This arrangement of rocks marks a 28-spoke wheel about 27 meters in diameter. Piles of rocks, called cairns, mark the center and six locations on the circumference. The astronomer John Eddy has discovered that these cairns mark sight lines toward a number of important points on the eastern horizon and could have been used as a calendar to help schedule hunting, planting, harvesting, and celebrations.

Many other American Indian sites have astronomical alignments. More than three dozen medicine wheels are known, although most do not have the sophistication of the Big Horn Medicine Wheel. The Moose Mountain Wheel, for instance, seems to have been in use as early as A.D. 100. Alignments at some Mound Builder sites of the Midwest show that these peoples were familiar with the sky. In the Southwest, the Pueblo Indian ruins in Chaco Canyon, New Mexico,

among others, have alignments that point toward the rising and setting of the sun at the summer and winter solstices.

Primitive astronomy also flourished in Central and South America. The Mayan and Aztec empires built many temples aligned with the solstice rising of the sun and with the extreme points where Venus rose and set. The Caracol temple in the Yucatán is a good example. It is a circular tower containing complicated passages and a spiral staircase (thus the name *Caracol*—"the snail's shell"). The tubelike windows at the top of the tower point toward the equinox sunset point, and the most northerly and most southerly setting points of Venus. Unfortunately, only about one-third of the tower top survives, so the directions of any other windows are forever lost.

Archaeoastronomers are uncovering the remains of ancient astronomical observatories around the world. Some temples in the jungles of Southeast Asia, for instance, are believed to have astronomical alignments.

Other scholars are looking not at temples, but at small artifacts from thousands of years ago. Scratches on certain bone and stone implements seem to follow a pattern and may be an attempt to keep a record of the phases of the moon. Some scientists contend that humanity's first attempts at writing were stimulated by a desire to record and predict lunar phases.

Archaeoastronomy is uncovering the earliest roots of astronomy and at the same time is revealing some of the first human efforts at systematic inquiry. The key lesson of archaeoastronomy is that humans don't have to be technologically sophisticated to admire and study the universe.

One thing about archaeoastronomy is especially sad. Although we are learning how primitive people observed the sky, we may never know what they thought about their universe. Many had no written language. In other cases, the written record has been lost. Dozens of beautiful Mayan manuscripts, for instance, were burned by Spanish missionaries who believed they were the work of Satan.

That is one reason why our history of astronomy really begins with the Greeks. Some of their manuscripts have survived, and we can discover what they thought about the shape and motion of the heavens.

BACKGROUND NOTES 2 B

The Foucault Pendulum

In 1851, the French physicist Jean Foucault provided an unambiguous demonstration that Earth rotates on its axis. Foucault suspended a 25-kilogram mass on a 60-meter wire from the domed ceiling of the Pantheon in Paris. He started the pendulum swinging and soon noticed that the direction of the pendulum's swing, the "plane" of the pendulum's motion, was changing. The pendulum continued to swing back and forth, but the direction in which the pendulum was swinging changed. In other words, the pendulum was turning.

Foucault knew that the only force acting on the pendulum was gravity, and this force could not cause the pendulum to turn. He concluded that the plane of the pendulum was not turning but that Earth was rotating under the pendulum. What appeared to be the shifting of the direction of pendulum's plane of motion was actually the room and Earth turning beneath the pendulum.

BACKGROUND NOTES 4

Eratosthenes' Calculation of Earth's Circumference

Eratosthenes was one of the ancient Greeks who first applied logic and reasoning to explain and understand the natural world. The method he used for determining the circumference of Earth is a classic example of scientific reasoning.

Eratosthenes was an astronomer. Around 200 B.C., he used simple rules of geometry to develop a method of determining the circumference of Earth and hence its radius. The method Eratosthenes used was as follows:

He observed that the sun was directly overhead and shone down deep vertical wells in the ancient city of Syene, Egypt, at noon on the first day of summer. He also measured that the sun was about 7 degrees south of the zenith at the same time when viewed from his home in Alexandria, several hundred miles north of Syene. Eratosthenes knew that 7 degrees was roughly one-fiftieth of a circle. By applying geometry, he showed that the distance between the cities of Alexandria and Syene was approximately one-fiftieth of the circumference of Earth. A multiple of that distance gave him the approximate circumference of Earth. (See diagram on page 90.)

In Eratosthenes' time, the distance between Alexandria and Syene was measured at 5,000 stadia. A stadium is an ancient Greek unit of distance, roughly equal to 1/6 kilometer. Although we are not certain of the *exact* length of the stadium, Eratosthenes' calculation (50 x 5,000 stadia = 250,000 stadia) produced a result for the circumference of Earth that is within 5 percent of today's accepted value.

The accepted value today for Earth's circumference as measured by orbiting spacecraft is 40,070 kilometers. When one considers that Eratosthenes made his calculation using only a single linear unit of measure (the stadium) and his observation of a natural phenomenon, his accomplishment is quite impressive. He

had no calculators, computers, or observational satellites.

You may be curious as to why Eratosthenes calculated the circumference of Earth as opposed to the length and width of a flat surface. Greek scientists had known that Earth was round since at least the fourth century B.C. They had observed that a ship's mast slowly sank below the horizon as the ship sailed out to sea. If Earth was flat, they reasoned, the mast would disappear abruptly. Furthermore, the shadow of Earth during eclipses of the moon was also observed to be round, thus indicating that the object producing the shadow was a sphere.

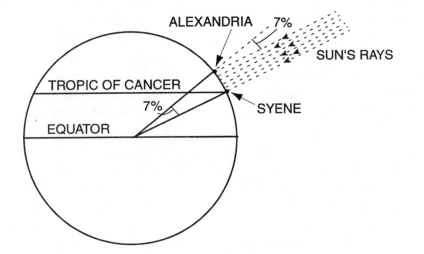

ERATOSTHENES' MEASUREMENT OF THE CIRCUMFERENCE OF EARTH
250,000 STADIA X 1/6 KILOMETER/STADIUM
= 42,000 km

The Law of Universal Gravitation

Newton's law of universal gravitation was a simple statement, based on his observation of the moon and of objects on Earth, that all objects attract all other objects. The force of that attraction is proportional to the masses of the objects and inversely proportional to the square of the distance between them.

When Newton put forth his law of universal gravitation in his *Principia*, other scientists wondered about how a force could act on objects at very great distances. Newton's concept of gravity seemed almost mystical, not scientific. Newton made no effort to explain what gravity was, only to state that the law worked. The law did work in describing how the moon orbited Earth and how objects fell to the surface of Earth. However, over the next hundred years, scientists began to use the law of gravity to describe the motion of objects at great distances and began to demonstrate that the law appeared to be universal.

The first such demonstration was conducted by Edmund Halley in the late 1600s. Studying astronomical records, Halley noticed that the orbits of several bright comets that appeared in 1531, 1607, and 1682 were similar. Because of the similarity of the orbits, Halley believed that these were appearances of a single comet. He used the newly published law of universal gravitation to predict the return of this comet. His prediction came true in 1758, 16 years after his death. We now know that this comet has been observed at every return since 240 B.C.; its most recent appearance was in 1985–1986. Comet Halley was named in honor of the first man to calculate its orbit.

Nearly 80 years after the first predicted return of Comet Halley, John Couch Adams and Urbain Leverrier used the law of universal gravitation to explain the peculiar motions of Uranus. Discovered accidentally by William Herschel in 1781, Uranus was observed not to follow exactly the orbit predicted by Newton's law of universal gravitation. Scientists understood that gravitational forces other than the sun acted on Uranus. These additional forces are referred to as gravitational perturbations. But even after taking into account the effects of the known planets, Uranus did not follow the orbit predicted by Newton's laws. Leverrier and Adams suggested that another planet beyond the orbit of Uranus was causing the extra perturbations. They independently calculated the position of the unknown planet. Their calculations proved to be correct when Neptune was discovered in 1846. Universal gravitation applies at distances of at least 30 astronomical units.

The first observation that the force of gravity worked beyond the solar system and between stars was made by William Herschel in 1804. He noticed that the fainter companion of the star Castor moved slightly relative to its brighter component. It required the application of Newton's law of universal gravitation to show that the gravity between stars was the same as the force between planets in our solar system. That proof came in 1827, when Felix Savary was able to show that the orbits of the stars in the binary system XI Ursae Majoris were ellipses. With the determination of the 60-year period of their orbits about each other, Newton's law of gravitation became truly universal.

Special and General Relativity

Introduction

These background notes are provided to help you understand the origin, postulates, and consequences of the special and general theories of relativity. One of the difficulties in understanding these laws is that, except for a few scientists, most people have never had any personal experience with the conditions that relativity explains. No aircraft or spacecraft has ever traveled fast enough for the consequences of special relativity to become significant. We have never been in a gravitational field so strong that our motion had to be described by general relativity. Because we lack personal experience with these theories, the results or consequences these theories predict cannot be arrived at by using our common sense. In other words, special and general relativity are counterintuitive because they seem to defy common sense. But, think again. Why should nature have to follow our expectations, which are based on our experiences, in areas in which we have never personally participated?

Although the laws of relativity are counterintuitive, it is possible to understand them. One way to learn these topics is to approach them from a historical point of view. The following notes will develop the history surrounding the evolution of the special and general theories of relativity. Through this history, we can see what problems arose during the study of certain events and hypotheses. We can learn what individuals were involved in the efforts and what contributions and missteps these individuals made. By following the experiments that have been conducted, we can see the physical evidence that supports the special and general theories of relativity. Although these theories require some mental gymnastics to understand, we can be confident that substantial experimental evidence supports these laws. Remember, in order for these theories to be useful, they must explain related events and predict new phenomena to observe.

Searching for the Ether

From the time of Aristotle through the end of the nineteenth century, scientists and philosophers wondered about the nature of the material that existed between Earth and the sun. They believed light needed something through which to travel in its journey from the sun to Earth. It was difficult, if not impossible, to conceive that light could travel such great distances through a void. Thus "something" had to exist in that space. Aristotle devised the notion of the ether, and its existence became a central component of nineteenth-century science. Its existence was made even more necessary when experiments were conducted in the 1840s and 1850s that proved that light had wavelike properties. The logic went something like this: If sound waves require air through which to travel and water waves require water through which to travel, then light waves must require a material (ether) through which to travel.

The existence of the ether was so well accepted that its properties were well defined. The speed of a wave depends on the stiffness of the medium through which it travels, and since light is the fastest of all waves, the ether must be extraordinarily stiff or simply incompressible. Furthermore, Earth was not slowed in its journey around the sun because of the ether. Therefore, the ether had no viscosity. It was a perfectly transparent fluid. Based on these properties, the only thing left to do was to detect the existence of the ether. The results of these experiments to detect the ether revealed cracks in Newton's classical laws of physics.

Michelson-Morley Experiment

The most famous and accurate of all the experiments to detect the ether was conducted by Albert A. Michelson and Edward R. Morley at the Case Institute of Technology in Cleveland, Ohio, in 1887. The experiment consisted of splitting a beam of light into two individual beams that traveled at right angles (perpendicular) to each other along the arms of an instrument called an interferometer. The arms of the interferometer were of exactly equal length and had a small mirror at each end. The beams of light reflected off these mirrors and returned, traveling exactly the same distance, into an eyepiece to be studied. The beams overlapped and created an interference pattern of light and dark circles. If the beams came back at exactly the same time, the central circle would be bright. If one beam trailed behind the other by the even smallest amount, the central circle would be dark. This instrument could detect the difference in arrival times of one part in 100 million.

Since Earth was believed to move through the ether, the hypothesis was that the beam of light moving in the direction of Earth's motion around the sun would be affected by the ether and lag behind the beam traveling perpendicular to Earth's motion. (The experiment was similar to a race between airplanes, in which one plane had to fly with both a very stiff head wind and a tail wind. The other plane, flying across the wind, would always win the race.) In the Michelson-Morley experiment, the expected results were a dark circle in the center of the interference pattern. The experiment was conducted thousands of times. The interferometer was rotated through a variety of angles and pointed in different directions. The results were always the same and opposite to what had been expected: a bright central circle in the pattern. The experiment had failed to detect the existence of the ether. Additional experiments conducted with better instruments over the next several decades all ended with the same results: The ether could not be detected. The conclusion of these findings was that the speed of light had to be unaffected by the motion of Earth. Thus, the speed of light was the same in all directions. Even though Michelson was the first American to win the Nobel Prize in physics, for his scientific instrumentation, he considered this experiment a failure and lamented it for the last 50 years of his life.

Numerous scientists attempted to explain the failure of the experiment. In 1892, G. F. Fitzgerald proposed that the interferometer arm that pointed in the direction of Earth's motion was contracted for some unknown reason. He actually developed an equation that related the amount of the contraction with the speed of the interferometer. Hendrik Lorentz, probably the most famous scientist at that time, believed very strongly that there must be a physical cause for the results of the experiment. He too accepted the notion that the interferometer was contracted. Lorentz believed that the contraction was due to some undetermined property of the electron, which had recently been discovered. Because all matter was made up of electrons, if the electrons were contracted in the direction of motion, so should be the interferometer.

Lorentz, however, went further. There was more to Michelson's experimental results than simply the contraction of the interferometer. He recognized that the speed of light had to be the same in all directions and for all observers. As a result, Lorentz developed the mathematics that showed not only how length contracted due to motion but also that time slowed for moving clocks. In other words, space and time were joined. It was no longer correct to think of absolute motion or absolute rest as Newton had done. The equations and theory developed by Lorentz were designed to explain the "failed" experiment of Michelson and Morley. The conclusion that the interferometer was shorter in one direction than in another seemed too obvious. The solution to the real problem was hinted at by the French mathematician Henri Poincaré in 1899, when he suggested that a new and more fundamental law was needed. He restated Galileo's principle of relativity: it is impossible to detect absolute motion or absolute rest in the laboratory. It was Albert Einstein who provided the fundamental law in 1905 by asking a completely different question.

Einstein's Special Theory of Relativity

Einstein observed an even more fundamental problem with Newtonian physics: an inconsistency between Newton's laws of motion and James Maxwell's (1864) laws of electricity and magnetism. In order to reconcile these two sets of laws, Einstein stated two fundamental propositions in his special theory of relativity. The first holds that the laws of physics are the same for all observers in uniform motion; the second states that the speed of light is the same for all observers regardless of their

motions. These two statements are enough to link the laws of electricity and magnetism with Newton's laws, explain the results of the Michelson-Morley experiment, and derive the Lorentz equations. The difference between the work of Lorentz and that of Einstein is that Lorentz was interested only in resolving the problem presented by the failure to detect the ether. The conclusion that the speed of light was the same for all observers was simply a consequence of the experimental result. On the other hand, Einstein was searching for an inner consistency between all the laws of physics; the fact that the speed of light was the same for all observers became a fundamental postulate.

It must be emphasized that the results or consequences of special relativity are just that, relative. Length contraction results in the shortening of moving objects only in the direction of their motion. Time dilation refers to the slowing of time (not merely clocks) for objects in motion relative to those clocks considered to be at rest. For example, length is contracted in a moving spacecraft by an observer who is considered to be at rest, and moving clocks run more slowly than clocks that are at rest relative to the observer.

One example of the consequences of special relativity is shown in the video program. The *muon* is a tiny particle that is created in Earth's upper atmosphere by bombarding cosmic rays. The particle is so unstable that it decays within one two-millionths of a second when produced in a laboratory. It would therefore be expected that few, if any, muons would reach Earth's surface. However, muons do reach Earth. Because of the muon's very high velocity, its "clock" runs more slowly than clocks on Earth. However, if you were to ride on a muon, the muon "clock" would, to you, appear to be running at normal speed. The muon disintegrates in one two-millionths of a second according to its clock. This sounds like a paradox. The solution depends on what the muon "sees." Its "clock" runs normally, but the muon "sees" Earth approaching it at very high velocity. The distance between the muon and the surface of Earth is contracted. The muon has less distance to travel and plenty of time to reach the surface of Earth. The muon illustrates the nature of special relativity.

Equivalence of Matter and Energy

Special relativity has several other consequences. The most relevant consequence of special relativity to the study of astronomy involves the equivalence of matter and energy. The famous equation $E = mc^2$ connects matter and energy and provides us with the understanding of how stars convert matter into energy.

The energy contained in matter is extremely large. The "c" in the equation refers to the speed of light. If you could convert 1 gram of matter (there are 28 grams per ounce), or roughly the mass of two paper clips, into total energy, you would have nearly 10^{14} joules of energy, or enough energy to supply your family's needs for more than 1,000 years. It is the conversion of matter into energy through the thermonuclear process of fusion that provides the sun with its energy. It is special relativity that provides us with the understanding of this process.

General Relativity

Special relativity is used to explain the behavior of objects that are traveling at speeds near the speed of light. Although relativity is a fundamental law that applies at all speeds, the effects are important only at high speeds. Furthermore, special relativity applies to constant speeds. Of course, not all speeds are constant. If the speed changes or if direction changes, there must be an acceleration. Einstein's general theory of relativity does more than simply extend special relativity to objects that accelerate, it provides a theory of the nature of gravity. The fundamental principle of general relativity is that acceleration is equivalent to gravity. In other words, the principle of equivalence states that it is impossible to distinguish between uniform acceleration and being at rest in a gravitational field.

Imagine the moment you step on your car's accelerator pedal. You are pressed back into the seat. Imagine the moment an elevator begins to rise upward. You feel somewhat heavier, you press a little harder on the floor. Imagine riding a roller coaster and reaching the bottom of the first drop. As you change direction and begin to move up the next rise, you feel yourself pressed hard into the seat of the roller coaster. All these sensations result from a change in velocity or acceleration. However, these sensations can also result from an increase in the gravitational force behind you. In other words, gravity can reproduce the same feelings you get when fuel is pumped into an engine or when your direction changes rapidly.

Acceleration and gravitation are equivalent. In fact, unless you saw the engine or felt the vibrations, it would be impossible to distinguish between gravitational effects and other forms of acceleration. This concept is the foundation of general relativity.

General Relativity and the Curvature of Space-Time

The study of special relativity has shown that the three dimensions of space (left and right, backward and forward, up and down) cannot be held separate from time, the fourth dimension. In fact, one cannot talk about an event unless it is located in space and time. We refer to this connection between space and time as *space-time*. Although space-time is fundamental to the nature of the universe, it is difficult to give a simple definition of space-time. Planets move through space-time. We conduct experiments in physics and astronomy in space-time. The galaxies move and collide in space-time. According to general relativity, the geometry of space-time is determined by the presence of matter. In other words, space-time is curved in the presence of matter.

If you were to fire a laser beam into space and aim it in such a way that it remains far from stars and other forms of matter, that beam of light would travel, for the most part, in a straight line. However, if the beam were to come near a large mass, that beam would have to follow along a curved path through space-time as produced by that large mass. Matter curves space-time, and matter and energy must follow along the paths of curved space-time. One way to look at curved space-time, although oversimplified, is to imagine a large flat sheet of rubber. Place a grapefruit in the middle of the sheet and the sheet bends in that area. Place a heavy sphere of iron and the sheet is even more curved. Roll a small object toward the depression in the sheet and that object will move along curved paths. The closer the object, the more curved the path. As long as the object remains on the rubber sheet, it must follow along those curved paths. (This analogy is two-dimensional, remember, space-time is four-dimensional.)

This concept of the curvature of space-time is very much different from Newton's view of gravity, that the gravitational force depends on the masses of the two objects involved. Light or other forms of energy have no "mass" and therefore, according to Newton's law of universal gravitation, should be unaffected by gravity. This is not the case. As first observed during a solar eclipse in 1919, beams of starlight were deflected around the sun, in accordance with the predictions made by general relativity. This experiment has been conducted numerous times with even greater degrees of accuracy.

Gravitational Red Shift/Time Dilation

The video program describes some of the other astronomical events explained by general relativity. The precession of the perihelion of Mercury's orbit and the bending of light around the sun are two of the most famous examples of experimental evidence that support and verify general relativity. Since gravity is said to curve space-time, then time must also be affected by gravity. Time runs more slowly in a stronger gravitational field. For example, if you place a clock at the top of a tall building and an identical one in the basement (the gravity at the top of the building is weaker than at the bottom because the top of the building is farther from the center of Earth), the clock in the basement will run more slowly than the clock on the top floor. In fact, an experiment with two atomic clocks, one at the top and one at the bottom of a tall building, has been conducted, and the results were consistent with those predicted by general relativity.

This slowing of time is an example of gravitational red shift, a phenomenon that has also been observed in the light emitted by massive stars as well as by the sun. Atoms within the sun that produce the light the sun emits can also be considered to be tiny clocks. Because the gravity of the sun and massive stars is large, the atoms emit light at a slightly lower frequency, or longer wavelength (slower vibrations) than the same atoms emitting light on Earth, which has a weaker gravitational field. Therefore, we would observe that the light from these atoms would be shifted slightly to the red end of the visible spectrum, hence the term gravitational red shift. Although difficult to measure, the slowing-of-time effect due to strong gravitational influences has been observed.

Gravity Waves

Another prediction of general relativity is the existence of gravity waves, or gravitational radiation. Imagine a boat moving through the water of a calm lake. The boat sets up waves that move through the water and may even reach the other side of the lake. By floating in the water, one could feel these waves and then realize that a boat has just passed by without even seeing the vessel. Massive objects do the same thing through space-time. As massive stars and other massive compact objects move through space-time, they create gravity waves that travel through space-time and produce ripples, just as a boat produces ripples on the water. It is theoretically possible to detect these waves and examine, in greater detail than visible light would allow, the nature of these objects. As of this date, however, gravity waves have not been directly detected.

Although gravity waves have not been directly detected, their existence has been implied by a number of experiments. The most famous experiment began in 1974 when Russell Hulse and Joseph Taylor discovered a binary (double) star system that consisted of two neutron stars, or pulsars. These stars, very dense and compact objects about ten miles across, emit pulses of radio energy at intervals so precise that they can be used as very accurate clocks. The period of the binary star system was measured to be 7 hours and 45 minutes. (An amazingly short

time, considering that it takes Earth one year to orbit the sun.) According to general relativity, the orbit of the binary system should be decaying or decreasing in size as the result of the emission of gravity waves. The first test of this hypothesis, reported in 1978, successfully confirmed that the system was decaying according to the predictions of general relativity. Additional experiments were conducted with even better results. In the final analysis, gravity waves have yet to be detected directly, but their existence is all but confirmed by the results of the experiments of Hulse and Taylor, who were honored for their work with the 1993 Nobel Prize in physics. As of the printing of this telecourse student guide in 2003, no instrument has detected gravitational radiation (gravity waves) directly. Yet that may change shortly. Two gravity "telescopes" have been designed and are under construction by the California Institute of Technology (Caltech) and the Massachusetts Institute of Technology (MIT).

The two telescopes are called LIGO, which stands for Laser Interferometer Gravitational-Wave Observatory. One telescope is located in Hanford, Washington, and the other is in Livingston, Louisiana. The two telescopes, located nearly 2,000 miles apart, are needed to confirm the same gravity wave event (a ripple in space-time) rather than some local disturbance.

Each telescope consists of two 4-foot-diameter vacuum pipes arranged in the shape of an "L" with 4-kilometer-long (2.5-mile-long) arms. At the crossing of the arms (the vertex of the L), and at each end, are masses suspended from wires and fitted with mirrors. Extremely stable lasers will traverse the pipes and reflect off the mirrors, meeting at the detectors and establishing an interference pattern.

As a gravity wave passes through the telescope, the spacing between these masses, and hence between the mirrors, will change. Furthermore, the change will be different from each arm of the "L." This change will cause the lasers to become "out-of-phase" and thereby create a new interference pattern. This new interference pattern will reveal the form of the passing gravitational wave. The large size of these telescopes will allow for the measurement of changes as small as 10^{-16} centimeters or one-hundred-millionth the diameter of the hydrogen atom.

Those interested in information about the construction and testing of LIGO may log on to the following website: www.ligo.caltech.edu/LIGO_web/about/factsheet.html or http://archive.ncsa.uiuc.edu/Cyberia/NumRel/LIGO.html. Return to the HOME page to learn about the current status and testing of LIGO.

General relativity is the best existing explanation for the nature of gravity and the behavior of objects that travel through space-time. General relativity not only describes the behavior of individual objects, but also the nature of the entire universe. It is the application of general relativity that makes it possible to describe the expanding universe (in fact, general relativity predicted the expansion, although Einstein refused to believe it) and to determine the requirements needed to predict its destiny.

7

Atoms and Starlight

Assignments

For the most effective study of this lesson, we suggest that you complete the assignments in the sequence listed below.

Before Viewing the Video Program

- Read the Overview and Learning Objectives for this lesson. Use the Learning Objectives to guide your reading, viewing, and thinking.

- Read Chapter 6, "Starlight and Atoms," pages 92–109, in the *Horizons* textbook.

- Read the Viewing Notes in this lesson.

View the "Atoms and Starlight" Video Program

After Viewing the Video Program

- Briefly note your answers to questions listed at the end of the Viewing Notes.

- Review all reading assignments for this lesson, especially the Chapter 6 summary on page 108 in *Horizons*.

- Respond to the challenges presented by each Critical Inquiry in the reading assignment and write brief answers to the review questions at the end of Chapter 6 in *Horizons* to be certain you understand the text material.

- Complete the Review Activities in this guide to reinforce your understanding of important terms and concepts. Check your answers with the Answer Key and review when necessary.

- Take the Self-Test in this guide to measure your achievement of the Objectives. Check your answers with the Answer Key and review when necessary.

- Complete any other activities and projects assigned by your instructor.

Overview

If the telescope was the most important invention to affect astronomy, then the spectrograph is certainly a close second. By using a wedged-shaped piece of glass known as a prism or a finely scratched mirror known as a grating, starlight can be dispersed into its component colors. Like spreading a hand of cards out in front of them, astronomers can spread out the light to see just what they have. Because light is the only "material" that distant stars provide, we must extract as much information from it as possible.

If the concepts in this lesson seem very abstract, especially for so early in the course, please understand that they are also very important. Since light is really what astronomers study, we must understand the nature of light and how it interacts with matter if we are to draw correct conclusions about the source of the light, whether that be a planet, star, nebula, galaxy, or even the flash of light that accompanied the birth of the universe.

From detailed analysis of a stellar spectrum, we can know what a star is made of, even though we have none of that star's material in our possession. The stellar spectrum also provides information about the temperature of the star's surface layers. The close similarity between the spectra of the sun and many other stars was one of the first indications that the sun *is* a star!

And, by using the Doppler effect (which is a change in the wavelength of radiation that results from the relative radial motion of the source of light and the observer of that light) astronomers can analyze spectra to reveal if a star is moving toward or away from Earth, and at what speed; if it is spinning slowly or rapidly; whether it is expanding or contracting, exploding or the target of infalling material; if its gases are calm or turbulent; and if there are actually two stars orbiting each other, even though we see only a single point of light. All this and more can be learned from the spectrum and, in most cases, there is no other way to obtain such information. This would be a much shorter course if the spectrograph had never been invented.

Learning Objectives

When you have completed all assignments for this lesson, you should be able to:

1. Describe a simplified model of a typical atom, with the arrangement of its subatomic particles and energy levels, and explain how such an atom interacts with light. *HORIZONS* TEXTBOOK PAGES 93–95; VIDEO PROGRAM.

2. Describe the appearance, origin, and characteristics of absorption, emission, and continuous spectra. *HORIZONS* TEXTBOOK PAGES 97–101; VIDEO PROGRAM.

3. Describe the origin of the first three series of lines in the spectrum of hydrogen. *HORIZONS* TEXTBOOK PAGES 100–101.

4. Explain spectral classification and how the appearance of a spectrum indicates stellar temperature. *HORIZONS* TEXTBOOK PAGES 99–105; VIDEO PROGRAM.

5. Explain what spectra indicate about stellar chemical composition and about stellar motions as revealed by the Doppler effect. *HORIZONS* TEXTBOOK PAGES 105–108; VIDEO PROGRAM.

Viewing Notes

The opening segment of the "Atoms and Starlight" video program provides historical perspective. It traces how the scientific study of light evolved from Newton's experiments with optics through the work of Joseph Fraunhofer, who discovered dark lines in the sun's spectrum in the early nineteenth century, to the work of Robert Bunsen and Gustav Kirchhoff, two German chemists who discovered the significance of those lines.

The segment describes the experiments Bunsen and Kirchhoff conducted that enabled them to identify the three basic types of spectra: continuous, emission, and absorption. They discovered that if you look through a spectroscope at the light from a heated, low-pressure gas (such as a solid powder that has been vaporized in a hot flame, or the electrified neon gas in a glass tube, or a beautiful nebula in space), you will see a bright-line spectrum. That is, the light will consist entirely of photons of a few specific colors, which will form a pattern of lines that is absolutely unique to the particular chemical that comprises the gas. But a heated, glowing material in the form of a solid, liquid, or high-pressure gas (such as the filament of a light bulb) will emit some light of *all* colors in a continuous spectrum. And, if we look at a glowing high-pressure gas through a thinner, cooler gas (as we do when we look at any normal star), we see a dark-line spectrum. That is, there is a nearly continuous "rainbow" with only certain colors missing.

When they compared the patterns of bright lines, which they obtained by throwing different chemicals into their Bunsen burner, to the patterns of dark lines which Fraunhofer had earlier spotted in the sun's spectrum, they saw that many of them were the same. Bunsen and Kirchhoff had discovered how to analyze the chemical composition of the sun and the stars. In the video program, computer graphics illustrate how each spectrum is produced and relate them to different celestial phenomena.

The next segment explains the essence of spectral analysis. We can tell what something is made of from the light that it emits because each chemical element produces a pattern of bright or dark spectral lines as unique as fingerprints. It has to do with the nature of matter and how it interacts with light. Within the atoms of which matter is composed, electrons orbit the nuclei. But unlike the planets orbiting the sun, electrons can only have specific orbits with specific energies. Only if a photon of light with just the right energy (and color) comes along, can the electron absorb that energy and jump to a higher orbit. And that electron can only emit a photon of a certain specific color and energy when it eventually falls back to a lower orbit. But since atoms of different chemicals have different sets of orbits, each different chemical will have a unique pattern of colors that it can absorb or emit.

In the segment "Spectroscopy" we look at modern spectrographs, the instruments which make such a detailed analysis possible. Two spectrographs featured in the program are the coudé spectrograph at Kitt Peak National Observatory and the fiber-optic spectrograph at Palomar Mountain Observatory. The coudé spectrograph is so large a person can walk around inside it, and the fiber-optic spectrograph can obtain the spectra of dozens of stars simultaneously by using fibers to position the light from each star along the spectrograph's slit. At both locations, observatory staff explain the operation of the instruments.

The sun and other normal stars are almost identical in their chemical composition, being mostly hydrogen and helium, with trace amounts of virtually every other chemical that exists in nature. Yet despite this uniformity, there are tremendous differences in the appearance of their dark-line spectra. As the final segment explains, this is because the outer, cooler layers of the stars vary substantially in temperature, and different chemicals will absorb light most efficiently at different temperatures. Astronomers themselves only came to this understanding about a century ago, after they had already begun to classify the spectra of stars by their varying appearance. Computer animation illustrates the modern spectral classification system—O, B, A, F, G, K, M—which you will be seeing so often throughout the course. The spectral classes were once more numerous and in alphabetical order. They were only rearranged into their present form after astronomers realized that surface temperature was the primary factor responsible for all the variety in appearance.

And yet the greatest value of the spectrograph may be in allowing us to measure the Doppler effect through slight shifts in the positions of a spectrum's dark or bright lines. The amount of these shifts is proportional to the speed with which an object is approaching or receding from us, and the direction of the shift tells us which way (blue shift for approach; red shift for recession). The video program concludes by explaining how, through the Doppler effect, a star's spectrum reveals clues about its motion with respect to Earth and its rotation rate. This same effect is used technologically in police radar, motion sensors, and such. We can use it to study the universe as well, and it may be the most versatile diagnostic tool which astronomers have at their disposal. You will be hearing references to Doppler shifts in almost every lesson of the telecourse!

As you watch the video program, consider the following questions:

1. What are the basic parts of an atom?

2. How does an electron absorb or emit a photon of light?

3. What are the three different kinds of spectra, and what kinds of objects (either astronomical phenomena or manufactured objects) produce them?

4. How and why can a spectrum be analyzed to determine temperature?

5. How and why can a spectrum be analyzed to determine chemical composition?

6. How and why can a spectrum be analyzed to determine a star's motion?

Review Activities

Matching

Match the terms listed below with the definitions that follow. Check your answers with the Answer Key and review any terms you missed.

_____ 1. nucleus

_____ 2. proton

_____ 3. neutron

_____ 4. electron

_____ 5. isotope

_____ 6. ionization

_____ 7. ion

_____ 8. molecule

_____ 9. Coulomb force

_____ 10. binding energy

_____ 11. permitted orbit

_____ 12. energy level

_____ 13. excited atom

_____ 14. ground state

_____ 15. black-body radiation

_____ 16. continuous spectrum

_____ 17. absorption line

_____ 18. absorption spectrum

_____ 19. emission line

_____ 20. emission spectrum

_____ 21. Kirchhoff's laws

_____ 22. transition

_____ 23. Lyman, Balmer, and Paschen series

_____ 24. spectral class

_____ 25. Doppler effect

_____ 26. radial velocity

a. rules governing the absorption and emission of light by matter

b. two or more atoms bonded together

c. energy emitted by a hypothetical perfect radiator

d. the kind of spectrum produced by a perfect radiator

e. the process by which atoms lose or gain electrons

f. the energy needed to pull a particular electron away from its atom

g. the kind of spectrum produced by a low-density gas

h. rate at which an object approaches or recedes from the observer

i. a dark-line in a spectrum

j. the central core of an atom

k. the state of an electron with a particular binding energy

l. any energy level that an electron may occupy within an atom

m. movement of an electron between two energy levels

n. electrostatic repulsion or attraction between charged bodies

o. an atom that has lost or gained one or more electrons

p. an indication of a star's surface temperature based on its spectrum

q. the lowest permitted energy level in an atom

r. an atom in which at least one electron is not in the ground state

s. some of the lines in the hydrogen spectrum

t. the shift in the wavelength of spectral lines due to radial velocity

u. atoms with the same number of protons but a different number of neutrons

v. a positively charged particle in the nucleus of an atom

w. low-mass atomic particle that carries a negative charge

x. atomic particle with no electrical charge

y. the kind of spectrum produced by the sun

z. a bright line in a spectrum

Completion

Fill each blank with the most appropriate term from the list for that paragraph. A term may be used once, more than once, or not at all. Check your answers with the Answer Key and review when necessary.

1. The spectrum produced by glowing high-density gas is a _____ spectrum, and its color is determined primarily by its _____. The spectrum produced by glowing low-density gas is a _____ spectrum, and its color is determined primarily by its _____. A normal star produces a _____ spectrum, its color is determined primarily by its _____, and the pattern of lines is determined primarily by its _____. The relative strengths of different lines in the pattern are determined primarily by _____.

   bright-line   dark-line

   chemical composition   temperature

   continuous

2. In the center of a model of a typical atom is a _____, which is surrounded by a cloud of _____. The two different particles that make up the center of the atom are the positively charged _____ and _____, which have no charge. The chemical element represented by an atom is determined by the number of _____. The isotope of the element it represents is determined by the number of _____. An ion is an atom in which the number of _____ is different from the number of protons.

   electrons   neutrons

   nucleus   protons

3. In a hydrogen atom, the higher the energy level, the _____ is the difference in energy between that energy level and an adjacent level. Thus, transitions to or from a higher level involve photons of _____ energies and _____ wavelengths than transitions to or from a lower level. In a very cool star, the number of hydrogen atoms with electrons in the second energy level would be _____ than those with electrons in the first energy level. In a very hot star, the number of hydrogen atoms with electrons above the

second level would be _____ than those with electrons in the second level. Thus, the cooler stars produce the strongest hydrogen lines in that part of the spectrum with _____ wavelengths.

lesser   shorter

longer   greater

Self-Test

Select the one best answer for each question. Check your answers with the Answer Key and review when necessary.

1. When an electron in an atom absorbs a photon with an energy less than the electron's binding energy, it may

   a. move to a lower energy level.

   b. remain in the same energy level.

   c. move to a higher energy level.

   d. escape from the atom altogether.

2. If left undisturbed by collisions or photons, an electron in an excited atom will

   a. eventually move to a lower energy level.

   b. remain in the same energy level forever.

   c. eventually move to a higher energy level.

   d. eventually escape from the atom altogether.

3. The hotter a star is, the

   a. brighter and redder it will be.

   b. brighter and bluer it will be.

   c. fainter and redder it will be.

   d. fainter and bluer it will be.

4. The sun produces
   a. a continuous spectrum.

   b. an emission spectrum.

   c. an absorption spectrum.

   d. no spectrum.

5. Which of the following is **NOT** one of the first three lines in the spectrum of hydrogen?
   a. Paschen series.

   b. Fraunhofer series.

   c. Lyman series.

   d. Balmer series.

6. Transitions to or from the ground state of a hydrogen atom produce spectral lines in the
   a. gamma ray part of the spectrum.

   b. ultraviolet part of the spectrum.

   c. infrared part of the spectrum.

   d. radio part of the spectrum.

7. A star's spectrum has very weak hydrogen lines but shows strong bands from titanium oxide. This star probably
   a. has very little hydrogen.

   b. is extremely hot.

   c. is moderate in temperature.

   d. is extremely cool.

8. Which one of the following spectral classes has stars with the coolest temperature?
   a. A

   b. B

   c. M

   d. O

9. The chemical element found in the greatest abundance in all normal stars is
   a. carbon.

   b. helium.

   c. hydrogen.

   d. iron.

10. Because of the Doppler effect, the spectral lines of a luminous object approaching Earth will appear
    a. broadened.

    b. split into several adjacent lines.

    c. shifted slightly toward the blue.

    d. shifted slightly toward the red.

Lesson Review

Lesson 7

Atoms and Starlight

Please Note: Use this matrix to guide your study and achieve the learning objectives of this lesson. It will also help you to view the video, which defines and demonstrates important concepts and skills as they relate to everyday life.

Learning Objective	Textbook	Telecourse Student Guide	Background Notes
1. Describe a simplified model of a typical atom, with the arrangement of its subatomic particles and energy levels, and explain how such an atom interacts with light.	pp. 93–95	Matching Activities: 1, 2, 3, 4, 5, 6, 7, 8, 9, 10, 11, 12, 13, 14; Completion Activities: 2; Self-Test Questions: 1, 2.	
2. Describe the appearance, origin, and characteristics of absorption, emission, and continuous spectra.	pp. 97–101	Matching Activities: 15, 16, 17, 18, 19, 20, 21; Completion Activities: 1; Self-Test Questions: 3, 4.	
3. Describe the origin of the first three series of lines in the spectrum of hydrogen.	pp. 100–101	Matching Activities: 22, 23; Completion Activities: 3; Self-Test Questions: 5, 6.	
4. Explain spectral classification and how the appearance of a spectrum indicates stellar temperature.	pp. 99–105	Matching Activities: 24; Self-Test Questions: 7, 8.	
5. Explain what spectra indicate about stellar chemical composition and about stellar motions as revealed by the Doppler effect.	pp. 105–108	Matching Activities: 25, 26; Self-Test Questions: 9, 10.	

8

The Sun

Assignments

For the most effective study of this lesson, we suggest that you complete the assignments in the sequence listed below.

Before Viewing the Video Program

- Read the Overview and Learning Objectives for this lesson. Use the Learning Objectives to guide your reading, viewing, and thinking.

- Read Chapter 7, "The Sun—Our Star," sections 7-1 and 7-2, pages 110–125, and the chapter summary, pages 129–130, in the *Horizons* textbook.

- Read Background Notes 8, "Solar Magnetic Cycle," following Lesson 13 in this guide.

- Read the Viewing Notes in this lesson.

View "The Sun" Video Program

After Viewing the Video Program

- Briefly note your answers to the questions listed at the end of the Viewing Notes.

- Review all reading assignments for this lesson, especially the Chapter 7 summary on pages 129–130 in *Horizons*, Background Notes 8 from Unit II, and the Viewing Notes in this lesson.

- Respond to the challenges presented by each Critical Inquiry in the reading assignment and write brief answers to review questions 1–11 at the end of Chapter 7 in *Horizons* to be certain you understand the text material.

- Complete the Review Activities in this guide to reinforce your understanding of important terms and concepts. Check your answers with the Answer Key and review when necessary.

- Take the Self-Test in this guide to measure your achievement of the Objectives. Check your answers with the Answer Key and review when necessary.

- Complete any other activities and projects assigned by your instructor.

Overview

So far in this course, we have seen the long historical process by which astronomy tore itself from its ancient mystical origins and became a modern science. We have viewed the ever more sophisticated telescopes, spacecraft, and auxiliary instruments that astronomers rely upon to study the universe. And we have learned some of the basic laws of physics, relating to light and gravity for example, which govern everything in nature, but are particularly important in understanding phenomena in outer space. Yet we have not really begun to explore the astronomical phenomena themselves.

In this lesson, and for the remainder of the course, we will do just that, beginning with the sun. It is a fitting object with which to start a journey that will carry us increasingly farther into space and, in some cases, farther back in time. For the sun is a star, very similar in most respects to those myriad points of light in the nighttime sky, but *vastly* closer to us than any other. It is the only star whose surface behavior can be studied in detail. Probing its secrets has certainly helped us to understand stars in general, but it works the other way too. For example, the presumption that the sun will eventually swell into a huge, intensely bright red giant star and destroy us (as described in Lesson 12), is based on our watching other, older stars undergo that very change, even though our sun has not done that . . . yet.

Although our dependence upon the sun's light and heat is obvious, we are slowly learning that the sun is curiously undependable. Its 11-year magnetic cycle, revealed most clearly by the changing number of sunspots, has long been known to wreak havoc with our more delicate technologies, such as navigational and communications equipment and orbiting satellites. That cycle is only poorly understood, so it is even more perplexing when the cycle shuts down for extended periods, as it has several times in the past few thousand years, apparently resulting in prolonged cold spells on Earth. For all of its nearness, the sun remains mysterious in ways that affect our lives.

This lesson concentrates on the sun's surface features and behavior, and their effect on Earth, the very kinds of phenomena that cannot be studied easily in distant stars. But the next five lessons are devoted to every other aspect of stellar birth, life, and death, and will include a more detailed description of the sun's hidden interior and nuclear reactions.

Learning Objectives

When you have completed all assignments for this lesson, you should be able to:

1. Identify and describe the three major layers of the solar atmosphere, how each is observed, and the principal phenomena located in each layer. *HORIZONS* TEXTBOOK PAGES 111–117; VIDEO PROGRAM.

2. Describe the behavior and suspected cause of the solar magnetic cycle and its effects on sunspots, prominences, flares, and coronal activity. *HORIZONS* TEXTBOOK PAGES 117–124; VIDEO PROGRAM.

3. Describe known and suspected effects of the solar magnetic cycle on Earth and its immediate environment in space. *HORIZONS* TEXTBOOK PAGES 119 AND 125; BACKGROUND NOTES 8; VIDEO PROGRAM.

4. Describe variations from the usual solar magnetic cycle and their suspected effects on Earth. *HORIZONS* TEXTBOOK PAGE 119; BACKGROUND NOTES 8; VIDEO PROGRAM.

Viewing Notes

After the introduction reminds us of the sun's commonness with other stars, the first segment emphasizes how differently we study the sun than other stars. It concerns solar telescopes and related tools. And although the ultimate goal of solar and stellar observations may be the same, to study spectral features for instance, the sun's blinding brightness and atmospheric turbulence-causing heat make the specifics of solar observing equipment quite unique.

The video program shows portions of the 60-foot solar tower on Mount Wilson which was used, along with the Zeeman effect, to discover the magnetic nature of sunspots early in the twentieth century. And you'll be introduced to helioseismology, which promises to use vibrations in the visible surface of the sun to reveal the underlying hidden structure, especially the sun's rate of rotation at different depths. The advantage of studying the sun from space is explained. It enables us to avoid the turbulence in Earth's atmosphere as well as atmospheric absorption of ultraviolet, X-ray, and other wavelengths we wish to study. It enables us to study solar features, like the corona, without having to wait for a total solar eclipse, and to sample the solar wind, which flows throughout the solar system. The program mentions the pioneering work of Skylab and the Solar Maximum Mission and previews the Ulysses mission, which was sent to study the sun from above its poles, a view not possible from Earth. Ulysses went to Jupiter first and used that planet's tremendous gravity to hurl the spacecraft above the plane of the solar system and back toward the sun.

The next segment uses computer graphics to provide a complete inside-out view of the sun's structure. Starting with the core at the very center where temperature and pressure is so great that nuclear reactions occur, we are then shown the radiative zone through which the photons of released energy flow smoothly, and the convective zone in which they don't, resulting in an overturning, boiling-like motion in the gas. These first three layers constitute the sun's *interior* and will not be described in detail until Lesson 11. They were included briefly in this video because the convection is largely responsible for what we see at the surface: granules and supergranules, sunspots and the solar wind, differential rotation and the magnetic cycle—all of which are important topics for this lesson. Above the convective zone is the solar atmosphere, consisting of the photosphere, which is the apparent visible surface, the chromosphere, and the corona. These are described in considerable detail in the text.

The "solar transition zone," which Dr. Rhodes indicates is located between the chromosphere and the corona, can be thought of as the innermost region of the corona, where its temperature rises most rapidly and dramatically. And when Dr. Zirin refers to "Voyager" still being stuck in the solar wind, he is speaking of the two Voyager spacecraft (which took almost all of the photographs in Chapter 18) that are now beyond the farthest planets and still detecting the wind. They hope to continue monitoring it until it merges into the interstellar gases at the end of the solar system—in a sense, the real edge of the sun.

The final segment of the video concerns the solar magnetic cycle and its effect on Earth and interplanetary space. It looks at sunspots, prominences, and flares in detail and uses the Babcock model to suggest that the ability of the gaseous sun to rotate at different speeds at different latitudes is somehow responsible for the 11-year magnetic cycle. But the Babcock model cannot account for the occasional cessation of that cycle and corresponding effects on Earth's climate. The video does not have time to explore the profound implications of this phenomenon (the experts describe the Maunder minimum briefly without even mentioning its name), so please see the additional Background Notes 8 following Lesson 13 in this guide. And understand that in his concluding remarks, Dr. Zirin is speaking beyond the Maunder-like anomalies every few centuries to consider the very long-term effects of the sun's aging process on itself and on us. Earth's environment has been remarkably stable (with liquid water oceans and at least primitive life-forms) even though the sun's brightness is thought to have increased some 30 percent over its, and our, 5-billion year history. But further brightening over the next few billion years is sure to do us in. Environmental consciousness can prolong our existence, but we can never survive the death of the sun. Lesson 12 will tell us why.

As you watch the video program, consider the following questions:

1. How does our study and knowledge of the sun differ from that of other stars?

2. What does the study of other stars tell us about our sun?

3. How is our sun studied? What are the advantages of studying the sun from above Earth's atmosphere?

4. What are the three layers of the solar atmosphere?

5. What features are found in the solar atmosphere?

6. What are some of the effects of the solar magnetic cycle?

Review Activities

Matching

Match the terms listed below with the definitions that follow. Check your answers with the Answer Key and review any terms you missed.

____ 1. auroras	____ 10. helioseismology
____ 2. Babcock model	____ 11. Maunder butterfly diagram
____ 3. chromosphere	____ 12. Maunder minimum
____ 4. corona	____ 13. photosphere
____ 5. coronal holes	____ 14. solar wind
____ 6. differential rotation	____ 15. spicule
____ 7. filtergram	____ 16. sunspot
____ 8. flare	____ 17. supergranule
____ 9. granulation	____ 18. Zeeman effect

a. the visible surface of the sun

b. a nearly invisible layer of gas exhibiting a brief flash of pink light during total solar eclipses

c. the solar atmosphere above the chromosphere

d. high-velocity atoms and ions streaming away from the sun

e. a cool, dark area of the sun's surface

f. the splitting of a single spectral line into multiple lines caused by a magnetic field

g. the pattern made by the distribution of sunspots over time

h. an extended period of reduced solar activity

i. varying rates of movement of specific parts of the same body

j. an explanation that the magnetic cycle results from progressive tangling of the solar magnetic field

k. a violent eruption on the solar surface

l. glowing displays in Earth's upper atmosphere caused by gusts in the solar wind

m. cooler, lower-density regions of the corona

n. photograph taken in the light of a specific region of the spectrum

o. mottled appearance of bright cells in the photosphere

p. study of the sun's modes of vibration

q. flamelike structures that can last from 5 to 15 minutes

r. a large convective feature on the sun's surface; spicules appear around the edges

Completion

Fill each blank with the most appropriate term from the list for that paragraph. A term may be used once, more than once, or not at all. Check your answers with the Answer Key and review when necessary.

1. The three major layers of the solar atmosphere are the _____, _____, and _____. The visible surface of the sun is the _____, which is a _____ layer of _____-density gas. The two layers that can only be seen during a _____ eclipse are the _____ and _____.

chromosphere	photosphere
corona	solar
high	thick
low	thin
lunar	

2. The sunspot cycle is related to the sun's _____. The sun, however, does not rotate uniformly. The equatorial region rotates _____ than regions at _____ latitudes. Moreover, the surface of the sun rotates _____ than the _____. These variations are known as _____ rotation and are clearly linked to the _____ cycle.

differential	lower
exterior	magnetic
faster	rotation
higher	slower
interior	

Self-Test

Select the one best answer for each question. Check your answers with the Answer Key and review when necessary.

1. The visible surface of the sun is called the
   a. photosphere.
   b. chromosphere.
   c. corona.
   d. heliosphere.

2. The highest temperature in the solar atmosphere is found in
   a. the photosphere.
   b. a sunspot.
   c. the chromosphere.
   d. the corona.

3. An absorption spectrum is produced by the
   a. photosphere.
   b. chromosphere.
   c. corona.
   d. solar wind.

4. The sun rotates
   a. fastest at the poles.

   b. fastest at mid-latitudes.

   c. fastest at the equator.

   d. equally fast at all latitudes.

5. The Zeeman effect is used to measure
   a. the rotation rate of the sun.

   b. vibrations coming from sun's interior.

   c. the speed of solar wind.

   d. the strength of solar magnetic fields.

6. The corona is largest and most circular
   a. at sunspot minimum.

   b. at sunspot maximum.

   c. at random times unrelated to sunspot cycle.

   d. during the Maunder minimum.

7. Displays of auroras are most frequent
   a. at sunspot minimum.

   b. at sunspot maximum.

   c. at random times unrelated to sunspot cycle.

   d. during the Maunder minimum.

8. Strong solar magnetic activity can have all of the following effects **EXCEPT**
   a. power outages.

   b. more gradual decay of spacecraft orbits.

   c. radiation hazard to astronauts.

   d. interference with radio communication.

9. The Maunder minimum lasted about
    a. 4 years.

    b. 11 years.

    c. 22 years.

    d. 70 years.

10. The Maunder minimum is thought to have
    a. caused unusually warm weather on Earth.

    b. caused unusually cool weather on Earth.

    c. increased auroral displays.

    d. had no discernible effect on Earth.

Lesson Review

Lesson 8

The Sun

Please Note: Use this matrix to guide your study and achieve the learning objectives of this lesson. It will also help you to view the video, which defines and demonstrates important concepts and skills as they relate to everyday life.

Learning Objective	Textbook	Telecourse Student Guide	Background Notes
1. Identify and describe the three major layers of the solar atmosphere, how each is observed, and the principal phenomena located in each layer.	pp. 111–117	Matching Activities: 3, 4, 7, 9, 10, 13, 14, 15, 17; Completion Activities: 1; Self-Test Questions: 1, 2, 3.	
2. Describe the behavior and suspected cause of the solar magnetic cycle and its effects on sunspots, prominences, flares, and coronal activity.	pp. 117–124	Matching Activities: 2, 6, 11, 16, 18; Completion Activities: 2; Self-Test Questions: 4, 5, 6.	
3. Describe known and suspected effects of the solar magnetic cycle on Earth and its immediate environment in space.	pp. 119, 125	Matching Activities: 1, 5, 8; Self-Test Questions: 7, 8.	Background Notes 8: Solar Magnetic Cycle
4. Describe variations from the usual solar magnetic cycle and their suspected effects on Earth.	p. 119	Matching Activities: 12; Self-Test Questions: 9, 10.	Background Notes 8: Solar Magnetic Cycle

Stellar Properties

Assignments

For the most effective study of this lesson, we suggest that you complete the assignments in the sequence listed below.

Before Viewing the Video Program

- Read the Overview and Learning Objectives for this lesson. Use the Learning Objectives to guide your reading, viewing, and thinking.

- Read Chapter 8, "The Family of Stars," pages 132–155, in the *Horizons* textbook.

- Read the Viewing Notes in this lesson.

View the "Stellar Properties" Video Program

After Viewing the Video Program

- Briefly note your answers to questions listed at the end of the Viewing Notes.

- Review all reading assignments for this lesson, especially the Chapter 8 summary on page 154 in *Horizons* and the Viewing Notes in this lesson.

- Respond to the challenges presented by each Critical Inquiry in the reading assignment and write brief answers to the review questions at the end of Chapter 8 in *Horizons* to be certain you understand the text material.

- Complete the Review Activities in this guide to reinforce your understanding of important terms and concepts. Check your answers with the Answer Key and review when necessary.

- Take the Self-Test in this guide to measure your achievement of the Objectives. Check your answers with the Answer Key and review when necessary.

- Complete any other activities and projects assigned by your instructor.

Overview

The previous lesson concerned the sun—a typical star, yet the only star whose surface we can see and study in detail. This lesson is the first of five lessons that deal with all the other stars in the sky. In this lesson, we face the challenge of how to study stars that, even when viewed through the largest contemporary telescopes, remain unresolved points of light. We cannot see their surfaces; they do not appear as even tiny circular disks of measurable diameter. Because of their tremendous distance from Earth, they are too small to be magnified significantly.

What measurements can be made of a featureless point of light? In one sense, not many. We can measure the direction of that point, that is, the location of the star in the sky as well as any changes in that location over time. And we can measure how much light there is, either in total or by using a spectrograph to dissect the light into its component colors. That's it. But in a larger sense, as this lesson illustrates, we can use this information to deduce many properties of stars such as distance, luminosity, surface temperature, mass, and diameter. And from these, we have slowly pieced together the answers to the grander questions: how do stars form (Lesson 10); how do they work (Lesson 11); and how do they die (Lessons 12 and 13)?

Learning Objectives

When you have completed all assignments for this lesson, you should be able to:

1. Define stellar parallax and parsec and describe the use of parallax to determine stellar distances. *HORIZONS* TEXTBOOK PAGES 133–135; VIDEO PROGRAM.

2. Explain how distance, luminosity, and apparent magnitude are related. *HORIZONS* TEXTBOOK PAGES 136–138; VIDEO PROGRAM.

3. Describe the purpose and design of the Hertzsprung-Russell (H-R) diagram, including the locations of supergiant, giant, main-sequence, and white dwarf stars, and explain how the H-R diagram is used in luminosity classification. *HORIZONS* TEXTBOOK PAGES 138–142; VIDEO PROGRAM.

4. Describe the behavior and observation of visual, spectroscopic, and eclipsing binaries and explain how they are used to determine stellar masses and diameters. *HORIZONS* TEXTBOOK PAGES 142–149; VIDEO PROGRAM.

5. Describe the mass-luminosity relation and the abundance of stars of various spectral and luminosity classes. *HORIZONS* TEXTBOOK PAGES 149–154; VIDEO PROGRAM.

Viewing Notes

This video program makes it undeniably clear that trying to understand the distant stars is an enormously formidable challenge, but the opening segment explains how we began to understand them anyway. The first breakthrough came in the early nineteenth century when stellar parallax became measurable for the first time. In a process similar to the triangulation used by surveyors, annual changes in a star's position due to Earth's motion around the sun reveal the star's distance from Earth. Though stars can be millions of miles in diameter, they are trillions of miles away at the closest. No wonder they look like mere specks of light in any telescope. And the quest to determine stellar distances continues as the program describes the mission of HIPPARCOS (the HIgh Precision PARallax COllecting Satellite) to extend such observations to a hundred thousand stars within our galaxy.

Distance is not a property of the star itself but, as the next segment explains, it can be used along with measurement of a star's apparent brightness to calculate a star's luminosity, its actual energy output. And luminosity *is* a fundamental stellar property that offers clues as to just how powerful a star's energy source must be. With some stars generating billions of times more energy every second than others, even though all of these stars are the same phenomenon in the same phase of their lives, with the same composition and the same energy source, we had a critical fact that any successful theory of stellar structure would have to be able to explain. To do that, we would need more clues.

The next segment reminds us how a star's surface temperature can be estimated from the pattern of its spectral lines or from its overall color. But the real insight came when the astronomers Ejnar Hertzsprung and Henry Norris Russell compared the luminosities and temperatures of stars in their famous diagram. Along with later measurement of stellar diameters, the diagram made it clear that the sun is a star and that all stars follow the same rules as any other glowing objects. The tremendous range in their luminosities resulted from their substantial variation in temperature and diameter (surface area). But we had still to identify the single stellar property that was responsible for determining all of the others.

The final segment does that. By watching the two members of a binary star system orbit each other, we can determine the amount of gravity (and therefore mass) needed to keep them from flying apart. And in those cases where the two stars eclipse each other as seen from Earth, we can also determine their diameters and approximate shapes. Stars are indeed spherical like the sun; they don't have five points!

The discovery of the mass-luminosity relation for main sequence stars turned out to be the crucial link. As Lesson 10 will explain, stars condense from clouds of interstellar gas, but some stars can end up more massive than others. And Lesson 11 will show that the more massive stars must also be hotter and brighter and bluer. Since they have more inward weight to support, they must balance that with the pressure of hotter, faster nuclear reactions in their cores. Mass turns out to be the singular defining characteristic that predestines a star's life story more directly and exactly than genes influence a human.

As you watch the video program, consider the following questions:

1. How does the parallax of an object depend on its distance from the observer as well as the distance between the two points from which the observation is made (the "baseline")?

2. How are a star's apparent brightness, luminosity, and distance related?

3. What are two ways to estimate the surface temperature of a star?

4. What two stellar properties are compared in a Hertzsprung-Russell diagram, and what is the relationship between these properties for main-sequence stars?

5. What third property can cause two stars of the same temperature to have different luminosities?

6. How are mass and luminosity related for main-sequence stars?

Review Activities

Matching

Match the terms listed below with the definitions that follow. Check your answers with the Answer Key and review any terms you missed.

_____ 1. stellar parallax

_____ 2. parsec

_____ 3. luminosity

_____ 4. H-R diagram

_____ 5. main sequence

_____ 6. giant star

_____ 7. supergiant star

_____ 8. white dwarf star

_____ 9. luminosity class

_____ 10. binary star

_____ 11. center of mass

_____ 12. visual binary

_____ 13. spectroscopic binary

_____ 14. eclipsing binary

_____ 15. light curve

_____ 16. mass-luminosity relation

_____ 17. absolute visual magnitude

_____ 18. spectroscopic parallax

a. two stars orbiting each other

b. a graph comparing the luminosities and surface temperatures of stars

c. the more massive a main-sequence star is, the more luminous it is

d. the angular difference in a star's position as seen from two different locations

e. a binary star system in which the two stars are not resolvable but exhibit variations in brightness as seen from Earth

f. the region on the H-R diagram occupied by about 90 percent of all stars and for which luminosity and surface temperature are most simply related

g. the point about which two stars move as they orbit each other, proportionally closer to the more massive star

h. the distance at which a star would have a parallax of 1 second of arc

i. the total amount of energy a star radiates each second

j. a binary star system in which both stars are visible separately

k. the intrinsic visual brightness of a star

l. a star about the size of Earth that is very faint though very hot

m. a category of stars determined by the widths of lines in their spectra

n. a graph indicating any change of brightness over time

o. a highly luminous star with a diameter about 10 to 100 times larger than that of the sun

p. an exceptionally luminous star with a diameter up to 1,000 times larger than that of the sun

q. a binary star system in which stars are too close together to be visible separately; presence of two stars is revealed by Doppler shifts

r. a method of estimating the distance of a star by comparing its apparent magnitude with its absolute magnitude as estimated from its spectrum

Completion

Fill each blank with the most appropriate term from the list for that paragraph. A term may be used once, more than once, or not at all. Check your answers with the Answer Key and review when necessary.

1. The farther a star is from Earth, the _____ its parallax. If we could observe the stars from Mars, which is farther from the sun than Earth, all of the stars would have _____ parallaxes. How bright a star appears depends on its distance and _____.

age	luminosity
larger	smaller

2. The Hertzsprung-Russell diagram compares the intrinsic brightness of a star to its _____. For main-sequence stars, the _____ they are in surface temperature, the brighter they are as well. If two stars have the same diameter but one is intrinsically fainter, it must be _____ in surface temperature also.

cooler	hotter
diameter	surface temperature

3. The motions of binary stars can be analyzed to tell us the _____ of each star. If the stars are eclipsing, we can also often tell the _____ of each star. For main-sequence stars, the _____ the mass, the greater the luminosity. For main-sequence stars, the larger the mass, the _____ the number of such stars we find of that type in space.

diameter	mass
larger	smaller .
luminosity	

Self-Test

Select the one best answer for each question. Check your answers with the Answer Key and review when necessary.

1. A parsec is
   a. the distance that light travels in 1 second.
   b. the number of seconds in a year.
   c. the average distance between Earth and the sun.
   d. the distance of a star with a parallax of 1 second of arc.

2. A star has a parallax of 0.1 second of arc. How far is it in parsecs?
   a. 0.1
   b. 1
   c. 10
   d. 100

3. If you moved five times farther from a distant star, the star would appear to be
   a. 5 times brighter.
   b. 5 times fainter.
   c. 10 times fainter.
   d. 25 times fainter.

4. Two stars are the same distance from Earth but star "A" has an apparent magnitude of 1 and star "B" has an apparent magnitude of 6. Which of the following statements is correct?
   a. Star "A" is 5 times more luminous than Star "B."
   b. Star "A" is 100 times more luminous than Star "B."
   c. Star "B" is 5 times more luminous than Star "A."
   d. Star "B" is 100 times more luminous than Star "A."

5. Of the following stars, the one with the largest diameter is
   a. G2 Ib.
   b. G2 II.
   c. G2 III.
   d. G2 IV.

6. Of the following main-sequence stars, the most luminous is
   a. B2.
   b. A4.
   c. F1.
   d. G8.

7. Stellar diameters can be determined by observation of
   a. astrometric binaries.
   b. eclipsing binaries.
   c. visual binaries.
   d. spectroscopic binaries.

8. For a binary star system with any given orbital size, the shorter the orbital period,
   a. the more massive the stars are.
   b. the less massive the stars are.
   c. the larger in diameter the stars are.
   d. none of the above; orbital period is unrelated to mass or diameter.

9. The rarest spectral class of stars is
   a. K stars.
   b. B stars.
   c. M stars.
   d. O stars.

10. The most massive main-sequence stars are
    a. A stars.
    b. G stars.
    c. O stars.
    d. M stars.

Lesson Review

Lesson 9
Stellar Properties

Please Note: Use this matrix to guide your study and achieve the learning objectives of this lesson. It will also help you to view the video, which defines and demonstrates important concepts and skills as they relate to everyday life.

Learning Objective	Textbook	Telecourse Student Guide	Background Notes
1. Define stellar parallax and parsec and describe the use of parallax to determine stellar distances.	pp. 133–135	Matching Activities: 1, 2; Completion Activities: 1; Self-Test Questions: 1, 2.	
2. Explain how distance, luminosity, and apparent magnitude are related.	pp. 136–138	Matching Activities: 3, 17; Completion Activities: 1; Self-Test Questions: 3, 4.	
3. Describe the purpose and design of the Hertzsprung-Russell (H-R) diagram, including the locations of supergiant, giant, main-sequence, and white dwarf stars, and explain how the H-R diagram is used in luminosity classification.	pp. 138–142	Matching Activities: 4, 5, 6, 7, 8, 9, 18; Completion Activities: 2; Self-Test Questions: 5, 6.	
4. Describe the behavior and observation of visual, spectroscopic, and eclipsing binaries and explain how they are used to determine stellar masses and diameters.	pp. 142–149	Matching Activities: 10, 11, 12, 13, 14, 15; Completion Activities: 3; Self-Test Questions: 7, 8.	
5. Describe the mass-luminosity relation and the abundance of stars of various spectral and luminosity classes.	pp. 149–154	Matching Activities: 16; Completion Activities: 3; Self-Test Questions: 9, 10.	

Stellar Formation

Assignments

For the most effective study of this lesson, we suggest that you complete the assignments in the sequence listed below.

Before Viewing the Video Program

- Read the Overview and Learning Objectives for this lesson. Use the Learning Objectives to guide your reading, viewing, and thinking.

- Read Chapter 9, "The Formation and Structure of Stars," section 9-1, pages 157–165, and section 9-5, pages 176–179, in the *Horizons* textbook.

- Read the Viewing Notes in this lesson.

View the "Stellar Formation" Video Program

After Viewing the Video Program

- Briefly note your answers to questions listed at the end of the Viewing Notes.

- Review all reading assignments for this lesson, especially the first three paragraphs of the Chapter 9 summary on pages 177 and 180 in *Horizons* and the Viewing Notes in this lesson.

- Respond to the challenges presented by each Critical Inquiry in the reading assignment and write brief answers to review questions 1–6 and 14 at the end of Chapter 9 in *Horizons* to be certain you understand the text material.

- Complete the Review Activities in this guide to reinforce your understanding of important terms and concepts. Check your answers with the Answer Key and review when necessary.

- Take the Self-Test in this guide to measure your achievement of the Objectives. Check your answers with the Answer Key and review when necessary.

- Complete any other activities and projects assigned by your instructor.

Overview

The previous lesson showed how astronomers can force stars to yield their secrets, despite their being incredible distances from Earth, in the guise of mere twinkling points of light. This lesson begins to describe the fruits of the astronomers' efforts. It concerns the formation of stars, the first step in the stellar life cycle that will be completed in Lessons 11 to 13.

We know that stars are huge spheres of mostly hydrogen and helium gas held together by their gravity and shining from the nuclear reactions taking place deep within them. Not surprisingly, stars first form from the interstellar medium—the even larger, though colder and thinner, clouds of hydrogen and helium found floating throughout our galaxy.

In trying to learn about stellar formation, astronomers have faced an enormous problem: they literally could not see the places where stars were being born. Because stars are so cool when first forming and are surrounded by thick clouds of dust that absorb whatever visible light the young stars may emit, the stars' birth process takes place in darkness, hidden from the view of ordinary telescopes. Astronomers, therefore, had to rely almost entirely on theory to detail the earliest phases of a star's life. But the onset of the space age and the introduction of orbiting infrared telescopes finally made it possible to peer through the dust, as ultrasound can reveal an unborn child in the human womb. Today, at nonvisible wavelengths, we can actually witness many aspects of stellar birth.

Learning Objectives

When you have completed all assignments for this lesson, you should be able to:

1. Describe the composition and distribution of the interstellar medium and how its contents are detected at visible and nonvisible wavelengths. *HORIZONS* TEXTBOOK PAGES 157–162; VIDEO PROGRAM.

2. Outline the theorized process by which stars form, from molecular cloud to main sequence. *HORIZONS* TEXTBOOK PAGES 162–164; VIDEO PROGRAM.

3. Describe confirming observations of actual pre-main-sequence objects at visible and nonvisible wavelengths. *HORIZONS* TEXTBOOK PAGES 165–167 AND 176–179; VIDEO PROGRAM.

Viewing Notes

The first segment of the video program "Stellar Formation" concerns the interstellar medium, the raw material of which stars are eventually made. Despite heated glowing clouds of gas and dark clouds of dust seen in silhouette against more distant fields of stars, earlier astronomers had to convince themselves that the space between the stars was not completely empty. The main evidence was the ability of the interstellar material to make fainter and redder the starlight passing through it. But today it is easy to spot this material. The warmth of dust appears brightly in infrared photographs, and the normally transparent hydrogen gas glows at the radio wavelength of 21 centimeters. The interstellar medium is primarily hydrogen and helium in gaseous form, mixed with smaller quantities of more complex molecules and fine, solid grains of dust such as graphite (like pencil "lead") and silicates (like beach sand). It is from this material that stars condense.

Under the right conditions, as the next segment describes, a cloud of gas and dust can become unstable and begin to collapse under its own weight, eventually fragmenting into many smaller clouds of varying mass. As each of these smaller fragments continues collapsing and heating until nuclear reactions ignite within it, an entire cluster of stars will be born all at once. Such a cluster is found within the Orion Nebula, a nearby region of our galaxy which is an especially prolific star factory, continuing to produce one generation of stars after another. And until the stars of a given cluster have a chance to separate, their proximity to each other leads to the captures and recaptures that result in certain kinds of binary star systems.

The many sophisticated techniques now available to astronomers to see through the interstellar dust are the focus of the concluding segment of the video program. Actually viewing most of the process of stellar birth had to await the introduction of infrared telescopes, which must be operated above as much of the water vapor and certain other gases in Earth's atmosphere as possible. Some infrared wavelengths can be observed at high altitude observatories or from balloons or high flying jets like the Kuiper Airborne Observatory, which is featured in the last segment of this program. Other wavelengths must be observed from space, as with the Infrared Astronomy Satellite (IRAS).

The last segment concerns observational confirmation of the theories outlined earlier. In the infrared, we can see the clusters of protostars embedded deep within the giant dark cold molecular clouds that shielded the stars from outside heating and enabled them to form. And in studying T Tauri stars and their associated variable nebulas, known as Herbig-Haro objects, we are actually witnessing the birth itself. As the young star's winds begin to blow, it rapidly clears away its surrounding cocoon of obscuring dust and emerges as another one of those visible twinkling points of light whose nature has so fascinated and, until recently, so eluded us.

As you watch the video program, consider the following questions:

1. How did astronomers come to conclude that dark nebulae were not holes in space but clouds of opaque dust?

2. What kinds of gas and dust are found in the interstellar medium? Under what circumstances can the interstellar medium be seen with optical telescopes, or with infrared or radio telescopes?

3. Why do stars form only from cold, dark clouds, and what kinds of external events can trigger such a cloud to form stars?

4. What is the significance of stars forming in clusters, such as in the Orion Nebula, and how can this lead to the formation of binary star systems?

5. What was the purpose of the Kuiper Airborne Observatory? (It has since been retired from service.)

6. In what ways can we actually observe protostars at different stages of their development? How do T Tauri stars and Herbig-Haro objects relate to star formation?

Review Activities

Matching

Match the terms listed below with the definitions that follow. Check your answers with the Answer Key and review any terms you missed.

____	1. interstellar medium		____	9. association
____	2. interstellar reddening		____	10. protostar
____	3. infrared cirrus		____	11. nebula
____	4. evolutionary track		____	12. birth line
____	5. emission nebula		____	13. T Tauri star
____	6. reflection nebula		____	14. Herbig-Haro object
____	7. molecular cloud		____	15. bipolar flow
____	8. shock wave		____	16. Bok globule

a. any visible cloud of gas or dust in space

b. a collapsing cloud of gas and dust destined to become a star

c. the gas and dust distributed between the stars

d. a dense interstellar gas cloud in which atoms can link together

e. the path a star follows on the Hertzsprung-Russell (H-R) diagram as it changes its surface temperature and luminosity

f. the process in which interstellar dust scatters away blue starlight

g. a cloud that glows because of starlight reflecting off interstellar dust

h. glowing interstellar gas cloud caused by ultraviolet light from hot stars

i. a sudden change in pressure that travels through space as a sound wave and can trigger star formation

j. the location on the H-R diagram where protostars of different masses first become visible

k. wispy network of cold interstellar dust clouds

l. group of widely scattered stars moving together through space but not gravitationally bound into a cluster

m. small, dark cloud of gas and dust believed to be about to form stars

n. young stars that seem to be clearing away surrounding gas and dust

o. the process by which T Tauri stars blow away surrounding material

p. small, variable nebula caused when a T Tauri star excites nearby interstellar gas

Completion

Fill each blank with the most appropriate term from the list for that paragraph. A term may be used once, more than once, or not at all. Check your answers with the Answer Key and review when necessary.

1. Most of the interstellar medium is in the form of _____, of which about 75 percent consists of the chemical element _____ and 25 percent of the element _____. Only about 1 percent of the mass of interstellar material is in the form of solid _____. This small amount of solid material creates the phenomenon of interstellar _____, which results from the fact that the longer wavelengths of _____ photons are _____ likely to be deflected than the shorter wavelengths of _____ photons.

blue	hydrogen
bluing	less
dust	more
gas	red
helium	reddening

2. Stars form from clouds that are _____ in temperature and _____ in density than normal. As the fragments of such a cloud continue collapsing to form stars, each fragment becomes steadily _____ in temperature and _____ in density until nuclear reactions begin and halt the collapse.

colder	hotter
greater	lesser

3. The more massive a protostar, the _____ amount of time it takes to form. During most of its existence, a protostar is detectable at _____ wavelengths, because it is _____ in temperature than normal stars and is also surrounded by _____. The T Tauri phenomenon is a protostar in the process of becoming _____. Herbig-Haro objects are clouds of glowing _____ lying on either side of a protostar, which is detectable at _____ wavelengths, or on either side of a T Tauri star, which is detectable at _____ wavelengths.

colder	infrared
dust	lesser
gas	nonvisible
greater	ultraviolet
hotter	visible

Self-Test

Select the one best answer for each question. Check your answers with the Answer Key and review when necessary.

1. The element that makes up most of the mass of the interstellar medium is
   a. hydrogen.
   b. helium.
   c. carbon.
   d. silicon.

2. Starlight passing through interstellar dust becomes
   a. fainter and bluer.
   b. brighter and bluer.
   c. fainter and redder.
   d. brighter and redder.

3. The wavelength at which the cold interstellar dust known as "cirrus" can be observed is
   a. gamma ray.
   b. ultraviolet.
   c. visible.
   d. infrared.

4. To begin to collapse into stars, an interstellar cloud must be
   a. cold and thin.
   b. cold and dense.
   c. hot and thin.
   d. hot and dense.

5. The dominant source of shock waves that trigger star formation in our galaxy is thought to be
   a. supernova explosions.
   b. ignition of very hot stars.
   c. collision of two interstellar clouds.
   d. the pattern of spiral arms.

6. A protostar of which of the following masses would take longest to finish forming?

    a. 5 solar masses

    b. 2 solar masses

    c. 1 solar mass

    d. 0.2 solar mass

7. Protostars are observable in the

    a. visible wavelength region.

    b. gamma ray wavelength region.

    c. infrared wavelength region.

    d. ultraviolet wavelength region.

8. Star formation is associated with all of the following objects EXCEPT

    a. red giants.

    b. molecular clouds.

    c. T Tauri stars.

    d. Bok globules.

9. The Herbig-Haro objects associated with any one protostar are frequently seen to occur in groups of

    a. two.

    b. four.

    c. a few dozen.

    d. hundreds.

10. One site of very active star formation is

    a. Crab Nebula.

    b. Orion Nebula.

    c. Cygnus X-1.

    d. Nova Cygni 1975.

Lesson Review

Lesson 10

Stellar Formation

Please Note: Use this matrix to guide your study and achieve the learning objectives of this lesson. It will also help you to view the video, which defines and demonstrates important concepts and skills as they relate to everyday life.

Learning Objective	Textbook	Telecourse Student Guide	Background Notes
1. Describe the composition and distribution of the interstellar medium and how its contents are detected at visible and nonvisible wavelengths.	pp. 157–162	Matching Activities: 1, 2, 3, 5, 6, 11; Completion Activities: 1; Self-Test Questions: 1, 2, 3.	
2. Outline the theorized process by which stars form, from molecular cloud to main sequence.	pp. 162–164	Matching Activities: 4, 7, 8, 9, 10; Completion Activities: 2; Self-Test Questions: 4, 5, 6.	
3. Describe confirming observations of actual pre-main-sequence objects at visible and nonvisible wavelengths.	pp. 165–167, 176–179	Matching Activities: 12, 13, 14, 15, 16; Completion Activities: 3; Self-Test Questions: 7, 8, 9, 10.	

The Lives of Stars

Assignments

For the most effective study of this lesson, we suggest that you complete the assignments in the sequence listed below.

Before Viewing the Video Program

- Read the Overview and Learning Objectives for this lesson. Use the Learning Objectives to guide your reading, viewing, and thinking.

- Read Chapter 7, "The Sun—Our Star," section 7-3, pages 123–129.

- Read Chapter 9, "The Formation and Structure of Stars," sections 9-2 through 9-4, pages 165–176, and the chapter summary, pages 177 and 180, in the *Horizons* textbook.

- Read the Viewing Notes in this lesson.

View "The Lives of Stars" Video Program

After Viewing the Video Program

- Briefly note your answers to questions listed at the end of the Viewing Notes.

- Review all reading assignments for this lesson, especially the Chapter 9 summary on pages 177 and 180 in *Horizons* and the Viewing Notes in this lesson.

- Respond to the challenges presented by each Critical Inquiry in the reading assignment and write brief answers to review questions 12–15 at the end of Chapter 7 and questions 7–13 at the end of Chapter 9 in *Horizons* to be certain you understand the text material.

- Complete the Review Activities in this guide to reinforce your understanding of important terms and concepts. Check your answers with the Answer Key and review when necessary.

- Take the Self-Test in this guide to measure your achievement of the Objectives. Check your answers with the Answer Key and review when necessary.

- Complete any other activities and projects assigned by your instructor.

Overview

The last lesson described the formation of a star as a process by which a cloud of interstellar gas collapses under its own weight, becoming denser and hotter at its core until "something" halts the collapse and stabilizes the star. That "something" is thermonuclear fusion, a variety of nuclear reactions capable of generating tremendous energy from the hydrogen gas that comprises the bulk of all normal stars. It is when the pressure from these reactions first balances a star's weight that the main-sequence phase of any star's life begins, a lengthy phase where the star changes very little until it eventually runs out of fuel and begins to die.

This lesson is about stars in their main-sequence phase, the most numerous kind, suspended between birth and death. But if previous lessons have made it clear how difficult it is to learn the external properties of stars like luminosity, surface temperature, mass, and diameter, how can we ever learn about their internal properties hidden deep inside? It's simple physics, really. The laws of stellar structure are based on observation of nature and common sense. With these laws and knowledge of a star's external characteristics, computers can create stellar models that detail the star's internal structure at any given distance from the center, including whether or not nuclear reactions are occurring in a particular layer. And a sequence of these models can predict how a star will change as it ages.

These models, painstakingly compared to observations of real stars at various stages of their lives, finally explain the reasons why massive stars are more luminous but also more scarce—something that was first revealed, but left unexplained, in Lesson 9.

Learning Objectives

When you have completed all assignments for this lesson, you should be able to:

1. Describe thermonuclear fusion, including the proton-proton chain, CNO cycle, and triple-alpha process, and explain how these reactions are governed by a pressure-temperature thermostat. *HORIZONS* TEXTBOOK PAGES 123–127 AND 165–169; VIDEO PROGRAM.

2. Describe the solar neutrino problem along with its possible explanations and implications. *HORIZONS* TEXTBOOK PAGES 127–129; VIDEO PROGRAM.

3. State the four laws of stellar structure and describe how they can be used to create a model of a star's interior. *HORIZONS* TEXTBOOK PAGES 169–173.

4. Explain the changes a star undergoes during the main-sequence phase and how the amount of time it spends in that phase is determined. *HORIZONS* TEXTBOOK PAGES 173–176; VIDEO PROGRAM.

5. Describe the theoretical explanations of the mass-luminosity relation and the abundance of stars of various spectral and luminosity classes as consequences of the nature of star formation and of the main sequence. *HORIZONS* TEXTBOOK PAGES 173–176; VIDEO PROGRAM.

Viewing Notes

The first segment of "The Lives of Stars" looks at the nuclear reactions themselves, the lifeblood of stars and the process that halts the gravitational collapse needed to give the star birth. Although there are many reactions detailed in the text, this video lesson concentrates on the proton-proton chain of nuclear reactions, a form of nuclear fusion, which produces most of the energy for the less massive stars such as the sun. In a series of reactions, the hydrogen of a star is transformed to helium along with substantial energy. In fact, stars generate energy so effortlessly that an energy-hungry humanity dreams of harnessing fusion on Earth, but the segment concludes with a look at the formidable technological barriers to such an accomplishment.

How do we know that stars generate nuclear energy deep inside? Theory says they should, and the next segment "Stellar Evidence" describe how computers generate models of a star's interior that, in turn, can predict the star's surface properties. These predictions are validated through comparisons with observations of real stars. And we can actually observe the sun's reactions indirectly because of the neutrinos—particles produced by one of the steps in the proton-proton chain—which stream out of the sun instantly instead of taking hundreds of thousands of years to filter to the surface as light does. The segment describes existing and proposed neutrino detectors, but also highlights the fact that we had been detecting far fewer neutrinos than we were expecting. Instead of this being a problem with the sun's nuclear furnace, however, the eventual explanation actually involved the peculiar nature of neutrinos and may affect the future of the entire universe, as will be described in Lesson 18.

The last segment applies all of this theory to explain observations of actual stars. Astrophysicists explain how a star's life is a constant battle between pressure and gravity. The evolution of stars is clearly demonstrated on Hertzsprung-Russell (H-R) diagrams. During the main-sequence phase, nuclear reactions keep the star's gravity from crushing it, but the weight of the star's outer layers keeps the pressure of those reactions from blowing the star apart like a huge hydrogen bomb. The more massive stars must generate energy more quickly and be externally hotter and brighter in order to balance the greater weight. But these most massive stars must also consume themselves most quickly. They have hundreds of times more fuel than the least massive stars but fuse it billions of times faster. The segment also looks at brown dwarfs, extremely low mass objects that form like stars, but never become hot enough in their cores to initiate nuclear reactions. (Since the video was produced, many likely brown dwarfs have been observed.) And, in setting the stage for the next lesson, we also see that a star begins to die the moment it finishes being born; the changing chemical composition within even main-sequence stars produces changes that accelerate near the end of a star's life and cause it to swell into a giant—large and bright but with not much time left.

As you watch the video program, consider the following questions:

1. What is the fuel in the proton-proton chain? What kinds of particles are produced? How is the energy generated related to the mass of the helium? Why are intensely high temperatures and pressures needed for such reactions and why don't all the reactions happen in a single step?

2. What kinds of approaches have been taken to harness thermonuclear fusion on Earth and what are some of the difficulties?

3. What is the purpose of a stellar model? What do you need to make one and what does it tell you?

4. What are different ways to detect neutrinos? What were some of the explanations for the shortage of neutrinos detected from the sun? Since the video was produced, scientists have confirmed that neutrinos have a slight mass and are oscillating among three different forms!

5. Why are main-sequence stars so stable? Why are the most massive stars the most luminous and the shortest-lived?

6. What becomes of forming stars which are too massive? Not massive enough?

7. What happens when a main-sequence star runs out of hydrogen in its core? How do observations of star clusters verify this conclusion?

Review Activities

Matching

Match the terms listed below with the definitions that follow. Check your answers with the Answer Key and review any terms you missed.

_____ 1. strong force

_____ 2. weak force

_____ 3. nuclear fission

_____ 4. Coulomb barrier

_____ 5. proton-proton chain

_____ 6. CNO cycle

_____ 7. neutrino

_____ 8. triple-alpha process

_____ 9. nuclear fusion

_____ 10. conservation of mass

_____ 11. conservation of energy

_____ 12. hydrostatic equilibrium

_____ 13. energy transport

_____ 14. opacity

_____ 15. stellar model

_____ 16. brown dwarf

_____ 17. lower edge of main sequence

a. the main, three-step energy source in the sun

b. the main, carbon catalyst energy source in more massive stars

c. the main energy source when a star becomes a red giant, with helium as the fuel

d. the process by which stars generate energy

e. the nuclear process used in uranium-fueled nuclear power plants

f. repulsion between nuclei overcome by violent collision

g. the reason it is difficult for electromagnetic radiation to escape a star

h. balance between material pressing down on a layer and pressure of gas in that layer

i. process by which hot objects cool

j. the reason there are no gaps within a star's layers

k. the reason a star's luminosity depends on its nuclear reaction rate

l. numerical description of conditions within each of a star's layers

m. nearly massless neutral atomic particle that travels at almost the speed of light

n. the natural force binding protons and neutrons together

o. the natural force responsible for some forms of radioactive decay

p. a star with too low a mass to ignite nuclear reactions

q. the location on an H-R diagram occupied by a newly formed star that has just stabilized

Completion

Fill each blank with the most appropriate term from the list for that paragraph. A term may be used once, more than once, or not at all. Check your answers with the Answer Key and review when necessary. If a question requires two or more answers in succession, they may be in any order, unless the question indicates otherwise.

1. In the proton-proton chain and CNO cycle, four _____ nuclei fuse to form a single nucleus of _____. In the CNO cycle, _____ is needed as a catalyst. In the triple-alpha process, the fuel is _____, and the principal byproduct is _____. The proton-proton chain dominates in the _____ massive main-sequence stars and the CNO cycle in the _____ massive stars.

carbon	less
helium	more
hydrogen	nitrogen

2. If a star is not generating enough energy to balance its weight, its diameter will _____, its central temperature will _____, and its reaction rate will _____, until it is in balance. The more massive a main-sequence star is, the _____ luminous it will be and the _____ time it will spend in the main-sequence phase. These massive main-sequence stars are _____ numerous than other types of stars.

decrease	less
increase	more

3. Because of hydrostatic equilibrium, as you go from the surface of a star to its center, its pressure will _____ and its temperature will _____. The three ways in which energy flows from hot regions to cooler regions are _____, _____, and _____. Close contact between molecules is necessary for _____ to occur. This method of energy transport is unimportant for most stars because their matter is _____. In the sun's inner regions, the principal means of energy transport is _____. In this type of energy flow, resistance to the flow is a function of the opacity of the gas. Heat builds up when the opacity of the gas is _____, creating a churning motion known as _____. This churning motion is important because it tends to make the gas _____.

conduction	increase
convection	low
decrease	radiation
gaseous	uniform
high	unstable

4. A brown dwarf is _____ massive than necessary to ignite nuclear reactions; over time, its temperature will _____.

decrease	less
increase	more

Self-Test

Select the one best answer for each question. Check your answers with the Answer Key and review when necessary.

1. In a main-sequence star, the number of hydrogen nuclei necessary to form a single helium nucleus via nuclear fusion is

   a. one.

   b. two.

   c. three.

   d. four.

2. The percentage of the mass of hydrogen that is converted to energy in nuclear fusion is approximately

   a. 100 percent.

   b. 99 percent.

   c. 50 percent.

   d. 1 percent.

3. The first neutrino detector to be built contained

   a. heavy water.

   b. chlorine.

   c. helium.

   d. gallium.

4. The number of solar neutrinos first detected by scientists was originally about

   a. 100 times as many as expected.

   b. twice as many as expected.

   c. the number expected.

   d. a third as many as expected.

5. If you hold the bowl of a spoon in a flame, the handle grows warm because of

    a. conduction.

    b. radiation.

    c. convection.

    d. fusion.

6. "The weight on each layer is balanced by the pressure in that layer" is the law of

    a. conservation of mass.

    b. conservation of energy.

    c. hydrostatic equilibrium.

    d. energy transport.

7. During the main-sequence phase, stars gradually become

    a. fainter and cooler.

    b. fainter and hotter.

    c. brighter and hotter.

    d. brighter and cooler.

8. A main-sequence star is twice as massive as the sun and 12 times more luminous. Its life span will be

    a. the same as the sun's.

    b. six times longer than the sun's.

    c. twice as long as the sun's.

    d. six times shorter than the sun's.

9. In comparison with main-sequence stars, brown dwarfs are

    a. more massive, smaller, and brighter.

    b. more massive, smaller, and fainter.

    c. less massive, larger, and brighter.

    d. less massive, smaller, and fainter.

10. In comparison with high-mass stars, low-mass stars are

    a. more common because they form naturally in larger numbers and have longer life spans.

    b. less common because they form naturally in fewer numbers and have shorter life spans.

    c. less common, even though they form naturally in larger numbers, because they have shorter life spans.

    d. more common, even though they form naturally in fewer numbers, because they have longer life spans.

Lesson Review

Lesson 11

The Lives of Stars

Please Note: Use this matrix to guide your study and achieve the learning objectives of this lesson. It will also help you to view the video, which defines and demonstrates important concepts and skills as they relate to everyday life.

Learning Objective	Textbook	Telecourse Student Guide	Background Notes
1. Describe thermonuclear fusion, including the proton-proton chain, CNO cycle, and triple-alpha process, and explain how these reactions are governed by a pressure-temperature thermostat.	pp. 123–127, 165–169	Matching Activities: 1, 2, 3, 4, 5, 6, 7, 8, 9; Completion Activities: 1; Self-Test Questions: 1, 2.	
2. Describe the solar neutrino problem along with its possible explanations and implications.	pp. 127–129	Self-Test Questions: 3, 4.	
3. State the four laws of stellar structure and describe how they can be used to create a model of a star's interior.	pp. 169–173	Matching Activities: 10, 11, 12, 13, 14, 15, 17; Completion Activities: 3; Self-Test Questions: 5, 6.	
4. Explain the changes a star undergoes during the main-sequence phase and how the amount of time it spends in that phase is determined.	pp. 173–176	Matching Activities: 16; Completion Activities: 2, 4; Self-Test Questions: 7, 8.	
5. Describe the theoretical explanations of the mass-luminosity relation and the abundance of stars of various spectral and luminosity classes as consequences of the nature of star formation and of the main sequence.	pp. 173–176	Completion Activities: 2; Self-Test Questions: 9, 10.	

The Deaths of Stars

Assignments

For the most effective study of this lesson, we suggest that you complete the assignments in the sequence listed below.

Before Viewing the Video Program

- Read the Overview and Learning Objectives for this lesson. Use the Learning Objectives to guide your reading, viewing, and thinking.

- Read Chapter 10, "The Deaths of Stars," pages 182–206, in the *Horizons* textbook.

- Read the Viewing Notes in this lesson.

View "The Deaths of Stars" Video Program

After Viewing the Video Program

- Briefly note your answers to questions listed at the end of the Viewing Notes.

- Review all reading assignments for this lesson, especially the Chapter 10 summary on pages 204–205 in *Horizons* and the Viewing Notes in this lesson.

- Respond to the challenges presented by each Critical Inquiry in the reading assignment and write brief answers to the review questions at the end of Chapter 10 in *Horizons* to be certain you understand the text material.

- Complete the Review Activities in this guide to reinforce your understanding of important terms and concepts. Check your answers with the Answer Key and review when necessary.

- Take the Self-Test in this guide to measure your achievement of the Objectives. Check your answers with the Answer Key and review when necessary.

- Complete any other activities and projects assigned by your instructor.

Overview

The preceding lessons described how stars are born and live; this one explains how they die at the hands of gravity. The process is complicated by exactly how much gravity is available.

Stars come in different masses for which the inward force of gravity is correspondingly different. But gravity is the driving force behind a star's entire life cycle. It's what collapses a cloud to form a star in the first place, what holds the star together while it is shining, and what waits patiently for the star to run out of energy so it can crush the star into some tiny, dense heap. And so a star's mass determines how quickly it forms, how quickly and brightly it lives, and how quickly it dies. But it also determines how it dies.

So this lesson must divide the subject of stellar deaths into three parts, each of which depends on mass. The low-mass red dwarfs collapse directly into white dwarfs, which is described only in the text. But the video program looks at the fate of medium-mass stars such as our own sun, which eventually will expand into a giant before puffing off its outer layers as a planetary nebula and *then* collapsing into a white dwarf. The sun's demise (which, of course, will destroy any life then on Earth) will therefore be gradual. But if a medium-mass star has a close binary companion star, mass transfer can lead to a more violent death in a nova or Type I supernova explosion.

The most massive stars will self-destruct, even without the help of companions, in Type II supernovae, when their dense, iron cores suddenly collapse and then eject the stars' outer layers. This lesson looks at the part of a supernova that is blown outward, enriching the interstellar gases with the heavier chemicals which had begun forming deep inside the star during its life. It will be the next lesson, however, that will consider what could become of that collapsing iron core.

Learning Objectives

When you have completed all assignments for this lesson, you should be able to:

1. Describe the internal processes that cause an aging main-sequence star to expand into a giant. *HORIZONS* TEXTBOOK PAGES 183–187; VIDEO PROGRAM.

2. Describe the deaths of low-mass red dwarfs and of medium-mass stars, including the formation of planetary nebulae and white dwarfs. *HORIZONS* TEXTBOOK PAGES 187–195; VIDEO PROGRAM.

3. Describe the deaths of the most massive stars, including Type II supernovae and supernova remnants. *HORIZONS* TEXTBOOK PAGES 198–204; VIDEO PROGRAM.

4. Explain how Hertzsprung-Russell (H-R) diagrams for different star clusters can be used to reveal details of the process of stellar evolution. *HORIZONS* TEXTBOOK PAGES 187–189.

5. Describe the evolution of close binary stars, including possible effects of mass transfer such as novae and Type I supernovae. *HORIZONS* TEXTBOOK PAGES 195–198 AND 201.

Viewing Notes

"The Deaths of Stars" video program begins by looking at the fate of our own sun, which is typical of the fate of medium-mass stars in general. The eventual depletion of hydrogen in the sun's core, and the subsequent cessation of nuclear reactions there, will cause the now pure helium core to contract. But this contraction further heats the core as well as its surrounding layers in which nuclear reactions still occur. As these reactions speed up, their excess energy causes everything but the core to begin to expand, brighten, and cool off, that is, to become a red giant. Although the helium eventually begins to react and will give the sun a brief reprieve from its death process, the later build-up of carbon and oxygen within the helium will cause the sun to expand even larger. It is during this time that the sun will gently shed its outermost layers in a beautiful bubble of gas known, inappropriately, as a planetary nebula (it has nothing to do with planets). And the newly revealed carbon-oxygen core is a ready-made white dwarf that will cool and fade slowly over billions of years.

But, as the next segment explains, some stellar deaths are much more sudden. If a nearby companion star manages to dump some of its hydrogen-rich outer gases onto the hot, dense surface of a white dwarf, nuclear reactions can begin there suddenly and blow those surface layers explosively into space in what is known, also inappropriately, as a nova (which means "new," but this is an old star). As brilliant and rapid as this explosion is, however, these thin surface layers are not missed by the white dwarf and do no damage; the process will begin to repeat itself perhaps thousands of times. In sharp contrast, though, is a white dwarf whose largely carbon interior is so close to igniting carbon fusion that a little extra gas from a companion star is all it takes to push it over the brink. When the carbon ignites as a Type I supernova, the star is destroyed from the inside out, leaving nothing in its wake.

Even stranger is the Type II supernova. It results from the virtual implosion of the iron core of a very massive star. As the surrounding layers fall onto this tiny core and rebound, an outward moving shock wave sweeps all of the star's outer layers violently into space, leaving only the tiny core behind (to give us something to talk about in Lesson 13). But those outer layers are an integral part of the chemical evolution of the universe. The early universe contained only lightweight chemicals such as hydrogen and helium. The heavier chemicals which constitute planets such as Earth, as well as the living creatures on it, did not yet exist. They were forged from the lighter chemicals by the nuclear processes inside stars, and when some of those stars exploded, these essential heavier chemicals were strewn into the surrounding interstellar medium so that later generations of forming stars could contain them.

Nowhere was that demonstrated more clearly than in the famous supernova 1987A, which is the subject of the program's final segment. It was the first supernova visible to the naked eye in almost four centuries and the only one since the invention of the telescope. It marked the first time that we knew what the star was like immediately before its outburst, and the first direct evidence for a collapsing iron core in the form of a flood of neutrinos detected just before the light of the explosion reached us. And the light and gamma rays observed in the days after the outburst revealed the telltale signs of some of the nuclear processes, occurring during the explosion itself, which manufacture those precious heavier chemicals that make up such a large portion of Earth and the living things upon it—including you.

As you watch the video program, consider the following questions:

1. What is the sequence of events in the death of a star like our sun, and what observable phenomena are associated with these events?

2. What are three different ways a dying star can explode? What masses of stars are involved? Which explosions are the most destructive? Which explosions are the least destructive? Which require a binary companion? What is left behind, and what is ejected in each case?

3. Why was supernova 1987A historically significant, and what have we learned from studying it?

Review Activities

Matching

Match the terms listed below with the definitions that follow. Check your answers with the Answer Key and review any terms you missed.

_____ 1. helium flash

_____ 2. degenerate matter

_____ 3. planetary nebula

_____ 4. black dwarf

_____ 5. Chandrasekhar limit

_____ 6. supernova remnant

_____ 7. synchrotron radiation

_____ 8. turn-off point

_____ 9. Roche surface

_____ 10. inner Lagrangian point

_____ 11. angular momentum

_____ 12. accretion disk

_____ 13. nova

a. conservation of this causes an object to spin faster as it gets smaller

b. the maximum mass of a white dwarf

c. in a gas of this type, nearly all of the lower energy levels are occupied

d. what a white dwarf becomes when it cools

e. what forms around a white dwarf as matter falls toward it

f. the outer boundary of the volume of space controlled by a star's gravity in a binary system

g. mass flows through this when passing from one star in a binary system to another

h. an explosion resulting from the transfer of matter from a normal star through an accretion disk onto the surface of a white dwarf

i. what is produced by the expanding gas when explosive carbon fusion or a collapsing iron core occurs

j. an explosive event in the cores of some giant stars that does not result in a supernova

k. produced by rapidly moving electrons spiraling through magnetic fields

l. the point at which a cluster's stars move from the main sequence toward the red-giant region

m. an expanding shell of gas released nonexplosively by a dying star

Completion

Fill each blank with the most appropriate term from the list for that paragraph. A term may be used once, more than once, or not at all. Check your answers with the Answer Key and review when necessary. If a question requires two or more answers in succession, they may be in any order, unless the question indicates otherwise.

1. Stars of different masses die in different ways. A red dwarf will collapse directly into a _____. A star like the sun will become first a _____, then eject its outer layers in a _____, and finally collapse into a _____. If this dying star receives additional mass from a binary companion, it can experience a nondestructive explosion on its surface called a _____ or a destructive explosion at its center called a _____. A very massive star will first become a _____ and then, when its _____ core collapses, produce a _____.

brown dwarf

carbon

giant star

iron

nova

planetary nebula

supergiant star

supernova

white dwarf

2. A main-sequence star becomes a _____ when it runs out of _____ in its core so that its purely _____ core collapses. Eventually, the _____ can ignite explosively without destroying the star. These new reactions will create a core made largely of _____. If this _____ core ignites explosively, it could cause the star to become a _____.

carbon

giant star

helium

hydrogen

iron

nova

supernova

3. The age of a star cluster containing massive blue stars must be relatively _____. The age of a star cluster containing only giant stars and low-mass red stars must be relatively _____. A star cluster's turn-off point also indicates the cluster's age. The lower the temperature of a cluster's turn-off point, the _____ the cluster is. The higher the cluster's turn-off point, the _____ it is.

old

older

young

younger

Self-Test

Select the one best answer for each question. Check your answers with the Answer Key and review when necessary.

1. A main-sequence star's outer layers first begin to expand rapidly into a giant star while
   a. hydrogen is burning in the core.
   b. helium is burning in the core.
   c. hydrogen is burning in a shell around the core.
   d. helium is burning in a shell around the core.

2. Degenerate matter is responsible for all of the following EXCEPT
   a. a gas where pressure does not depend on temperature.
   b. the ejection of a planetary nebula.
   c. the helium flash.
   d. explosive carbon fusion.

3. Of the following types of stars, the one that has never been and can never be a giant star is a
   a. red dwarf.
   b. white dwarf.
   c. black dwarf.
   d. Type II supernova.

4. The shell of gas ejected slowly by a giant star is known as a
   a. supernova remnant.
   b. nova remnant.
   c. planetary nebula.
   d. reflection nebula.

5. A Type II supernova is triggered by the collapse of a star's
   a. hydrogen core.
   b. helium core.
   c. carbon core.
   d. iron core.

6. The most recent supernova visible on Earth to the naked eye occurred in
   a. 1987.
   b. 1604.
   c. 1572.
   d. 1054.

7. A turn-off point corresponding to which of the following spectral classes would indicate the youngest star cluster?
   a. B
   b. A
   c. F
   d. G

8. A main-sequence star of which of the following colors is most likely to be youngest?
   a. red
   b. orange
   c. yellow
   d. blue

9. An explosive, although nondestructive, event on the surface of a white dwarf that is part of a binary system results in a
   a. planetary nebula.
   b. Type I supernova.
   c. Type II supernova.
   d. nova.

10. In the binary star system known as Algol,
    a. both stars are main-sequence stars.
    b. both stars are giant stars.
    c. only the more massive star is a giant.
    d. only the less massive star is a giant.

Lesson Review

Lesson 12

The Death of Stars

Please Note: Use this matrix to guide your study and achieve the learning objectives of this lesson. It will also help you to view the video, which defines and demonstrates important concepts and skills as they relate to everyday life.

Learning Objective	Textbook	Telecourse Student Guide	Background Notes
1. Describe the internal processes that cause an aging main-sequence star to expand into a giant.	pp. 183–187	Matching Activities: 1, 2; Completion Activities: 2; Self-Test Questions: 1, 2.	
2. Describe the deaths of low-mass red dwarfs and of medium-mass stars, including the formation of planetary nebulae and white dwarfs.	pp. 187–195	Matching Activities: 3, 4, 5; Completion Activities: 1; Self-Test Questions: 3, 4.	
3. Describe the deaths of the most massive stars, including Type II supernovae and supernova remnants.	pp. 198–204	Matching Activities: 6, 7; Completion Activities: 1; Self-Test Questions: 5, 6.	
4. Explain how Hertzsprung-Russell (H-R) diagrams for different star clusters can be used to reveal details of the process of stellar evolution.	pp. 187–189	Matching Activities: 8; Completion Activities: 3; Self-Test Questions: 7, 8.	
5. Describe the evolution of close binary stars, including possible effects of mass transfer such as novae and Type I supernovae.	pp. 195–198, 201	Matching Activities: 9, 10, 11, 12, 13; Completion Activities: 1, 2; Self-Test Questions: 9, 10.	

13

Neutron Stars and Black Holes

Assignments

For the most effective study of this lesson, we suggest that you complete the assignments in the sequence listed below.

Before Viewing the Video Program

- Read the Overview and Learning Objectives for this lesson. Use the Learning Objectives to guide your reading, viewing, and thinking.

- Read Chapter 11, "Neutron Stars and Black Holes," pages 207–227, in the *Horizons* textbook.

- Read the Viewing Notes in this lesson.

View the "Neutron Stars and Black Holes" Video Program

After Viewing the Video Program

- Briefly note your answers to questions listed at the end of the Viewing Notes.

- Review all reading assignments for this lesson, especially the Chapter 11 summary on page 226 in *Horizons* and the Viewing Notes in this lesson.

- Respond to the challenges presented by each Critical Inquiry in the reading assignment and write brief answers to the review questions at the end of Chapter 11 in *Horizons* to be certain you understand the text material.

- Complete the Review Activities in this guide to reinforce your understanding of important terms and concepts. Check your answers with the Answer Key and review when necessary.

- Take the Self-Test in this guide to measure your achievement of the Objectives. Check your answers with the Answer Key and review when necessary.

- Complete any other activities and projects assigned by your instructor.

Overview

This lesson completes the discussion of the lives of the stars. The previous lesson indicated that the final fate of a single low- or medium-mass star is to collapse into a white dwarf and then gradually cool into a black dwarf at some distant, future time. Only if such a star has a nearby binary companion is it sometimes possible to generate an explosive death.

But the previous lesson also explained that the most massive stars can also undergo destructive supernova explosions when their iron cores suddenly collapse. You've already seen that the violently ejected outer layers form beautiful supernova remnants. Now, this lesson completes the story by describing what can happen to the collapsing iron cores. Depending on each one's mass, it can become either a neutron star or a black hole—bodies even smaller, denser, and stranger than a mere white dwarf.

Unlike white dwarfs, which have the courtesy to shine and can be observed and measured like any other star, neutron stars and black holes would not shine in the usual sense and would have to be detected more cleverly. In the case of neutron stars, the accidental discovery of pulsars in 1967 was the breakthrough. Pulsars, which are observed to emit rapid, regular bursts of radio (and sometimes visible) light, were determined to be rapidly spinning neutron stars. That theoretical link was cemented by the discovery of a pulsar in the Crab Nebula, a supernova remnant which resulted from a single event which the ancient Chinese just happened, fortuitously, to have recorded for us in A.D. 1054.

As for black holes, astronomers are not 100 percent certain that they've found even one, although there are several good candidates. In every case, the black hole is identified by the effects of its tremendous gravity on a companion star or other material nearby. Understand that this lesson concerns "stellar mass" black holes, that is, the kind that form from the death of a single massive star. You will very soon be hearing about the supermassive black holes which may lurk in the cores of giant galaxies and which may each be as massive as millions of stars combined. Indeed, that is the direction that the course takes next; the next lesson, by starting with our own galaxy, the Milky Way, will begin to look at how individual stars are gravitationally joined together into the assemblages known as galaxies.

Learning Objectives

When you have completed all assignments for this lesson, you should be able to:

1. Outline the theoretical formation of neutron stars and describe their properties. *HORIZONS* TEXTBOOK PAGES 208–209; VIDEO PROGRAM.

2. Recount the discovery of pulsars and their identification as rapidly spinning neutron stars. *HORIZONS* TEXTBOOK PAGES 209–210; VIDEO PROGRAM.

3. Describe the observed properties of pulsars and binary pulsars. *HORIZONS* TEXTBOOK PAGES 210–219; VIDEO PROGRAM.

4. Describe the theoretical formation of black holes and their properties. *HORIZONS* TEXTBOOK PAGES 219–221; VIDEO PROGRAM.

5. Recount attempts to observe black holes indirectly and list examples of black hole candidates. *HORIZONS* TEXTBOOK PAGES 221–226; VIDEO PROGRAM.

Viewing Notes

After a quick review of the concept of a supernova, the first segment of the video program describes neutron stars, which are one of the two types of compact objects that can be left behind by the explosive deaths of the most massive stars. What we know of neutron stars is really the result of solving a nine-century-old celestial detective story. That's because the map that the Chinese drew of the "guest star" of 1054 pinpoints it at the same location that is now occupied by the Crab Nebula. And that nebula is expanding measurably at a rate that shows it formed at the same time the supernova appeared.

Astronomers also surmised that the collapsing iron core that triggers a supernova may be more massive than the Chandrasekhar Limit, the maximum mass that can exist as a white dwarf. So, the core must collapse to an even denser state, something called a neutron star which, at only a few miles across, could never shine with any detectable brightness. Those were the clues: a messy spattering of gas from an exploded star and the theory that neutron stars existed but were too small and too faint to be detected by direct evidence.

But the next segment shows otherwise. It concerns pulsars, an observed phenomenon whose discovery by radio telescopes in 1967 was accidental and unanticipated, and whose nature was at first impossible to explain. These highly precise radio pulses from space were even thought to be of artificial origin for a while until further measurements pointed to some natural effect. Eventually, a lighthouse-like motion was proposed, but only something as tiny as a neutron star could spin fast enough to produce such rapid pulses and, unfortunately, none of the then known pulsars were in the same direction as any known supernova remnant.

So, how could pulsars really be rapidly spinning neutron stars if neutron stars are formed by supernovae that leave glowing nebulae behind?

Like witnesses at a trial, the astronomers in this program answer that question. The most compelling evidence was the discovery of a thirty-times-per-second radio pulsar in the Crab Nebula at the same location as a faint star, and the subsequent use of a stroboscope-like device to show that star was pulsing *visibly* at the same rate. Any criminologist would have been proud. But the testimony concludes that there is a maximum mass even for a neutron star, so what happens if a collapsing iron core exceeds that also?

The third segment tackles that question by describing the only known alternative to a white dwarf or neutron star—a black hole. This time, the inward weight of the star is so great that it collapses to zero volume; its gravity is so strong that not even light can escape. The featured example of how such a bizarre object could be detected is Cygnus X-1. This is an X-ray source resulting from a binary star system where one star is no longer shining normally and is too massive to be anything other than a black hole. The X rays are believed to arise as gases from the other star circulate around the black hole and heat up tremendously as they wait their turn to be pulled down the "drain."

Although we can detect the presence of compact, massive objects in this manner, the final video segment concerns those theoretical properties of black holes which we have not been able to verify by observation. It looks at the stationary "Schwarzschild" black holes, with their singularities and event horizons, and at the rotating "Kerr" black holes, with their additional ergospheres. It talks about the highly unlikely possibility of using a black hole as a timewarp or spacewarp, and the realistic conclusion that any ordinary spaceship would be ripped apart before it could get anywhere near such a warp. It even mentions the postulated gravity waves, which we are technologically incapable of detecting at the moment. This last segment is all theory. But there was a time when the existence of the galaxies, which will dominate the next five lessons, was all theory too.

As you watch the video program, consider the following questions:

1. What is the evidence that the Crab Nebula is the result of a supernova?

2. Why is there a maximum mass that a white dwarf can have, and what happens when that mass is exceeded?

3. What evidence do we have that pulsars are rapidly spinning neutron stars?

4. Why is it not surprising that most pulsars are not in supernova remnants and most supernova remnants do not contain pulsars?

5. Why is there a maximum mass that a neutron star can have, and what happens when that mass is exceeded?

6. How can we detect the presence of a black hole?

Review Activities

Matching

Match the terms listed below with the definitions that follow. Check your answers with the Answer Key and review any terms you missed.

____ 1. neutron star

____ 2. pulsar

____ 3. lighthouse model

____ 4. millisecond pulsar

____ 5. singularity

____ 6. black hole

____ 7. escape velocity

____ 8. event horizon

____ 9. Schwarzschild radius

____ 10. time dilation

____ 11. gravitational red shift

a. what can form from the collapse of the most massive stars

b. what can form from the collapse of a star a little too massive to leave a white dwarf behind

c. the detectable evidence of a rapidly spinning neutron star

d. neutron star of the very most rapid spin rate

e. the object of zero radius into which all the matter of a black hole is thought to fall

f. the proposed explanation of a pulsar

g. equals the speed of light at event horizon

h. radius of the event horizon around a black hole

i. the reason an object never appears to cross the event horizon when seen from a distance

j. the reason a normal telescope could not detect a light of any brightness if located very near the event horizon

k. the boundary within which no event is visible to an outside observer

Completion

Fill each blank with the most appropriate term from the list for that paragraph. A term may be used once, more than once, or not at all. Check your answers with the Answer Key and review when necessary.

1. The initial mass of a star influences its fate. A collapsed star with a maximum mass of about 1.4 solar masses is a _____. A collapsed star that has maximum mass of about two to three solar masses is a _____. A collapsed object more massive than two to three solar masses should become a _____. Of these collapsed stars, the one that may be visible as a pulsar is a _____, and the one that can be seen directly is a _____. The type of collapsed star, the existence of which can only be inferred from its gravitational effect on, for example, the material of a nearby binary companion, is a _____. Of the three types of stars, the only one not typically formed from a collapsing iron core is a _____.

    black hole   white dwarf

    neutron star

2. Pulsars were first detected through the emission of _____. The discovery of pulsars is thought to confirm the existence of _____. According to theory, neutron stars are left behind by _____, and pulsars have, in fact, been detected inside remnants of these explosions. As pulsars age, they are observed to spin gradually _____. When a binary companion star adds mass to a pulsar, the pulsar will spin _____.

    black holes   radio waves

    faster   slower

    gamma rays   supernovae

    neutron stars   white dwarfs

    nova   X rays

3. Black holes are theorized to form if the core of a star collapses and has more than about _____ solar masses. If you were to approach a black hole, as you neared the event horizon, you would see yourself falling _____. Someone watching you from a distance would see you

falling _____ as you neared the event horizon. Once within the event horizon, your speed would always be _____ than the speed you need to escape.

faster   slower

four to five   two to three

one to two

Self-Test

Select the one best answer for each question. Check your answers with the Answer Key and review when necessary.

1. A typical neutron star is about the same diameter as
   a. the sun.
   b. Earth.
   c. North America.
   d. a city.

2. According to theoretical models, which of the following does **NOT** describe neutron stars in general?
   a. Its surface would be nearly as hot as our sun's interior.
   b. It would spin rapidly on its axis.
   c. It would be composed entirely of low-density gases.
   d. Its magnetic field would be a billion times stronger than Earth's.

3. Pulsars were first detected at
   a. radio wavelengths.
   b. infrared wavelengths.
   c. visible wavelengths.
   d. ultraviolet wavelengths.

4. The first pulsar to be found within a supernova remnant was inside the
   a. Orion Nebula.
   b. Cygnus Loop.
   c. Crab Nebula.
   d. Ring Nebula.

5. The radiation from a pulsar is believed to come in
   a. a burst from the entire surface each time the star expands to maximum diameter.
   b. a burst from the entire surface at each moment the star is expanding most rapidly.
   c. two continuous beams located at the poles around which the star rotates.
   d. two continuous beams located at the magnetic poles.

6. Millisecond pulsars are believed to be
   a. young pulsars that have always been isolated.
   b. young pulsars that have, or had, close binary companions.
   c. old pulsars that have always been isolated.
   d. old pulsars that have, or had, close binary companions.

7. The Schwarzschild radius for a ten-solar-mass black hole would be
   a. 0.45 centimeters.
   b. 1.5 kilometers.
   c. 30 kilometers.
   d. 10,000 kilometers.

8. If you replaced the sun with a one solar mass black hole, the gravity holding Earth in orbit would become
   a. infinite.
   b. stronger, and Earth would be pulled in.
   c. no different than it is now.
   d. weaker, and Earth would escape into space.

9. Which of the following objects is a likely candidate for containing a black hole?

    a. Hercules X-1

    b. Cygnus X-1

    c. Crab Nebula

    d. Orion Nebula

10. Black hole candidates are conspicuous by their continuous emission of

    a. gamma rays.

    b. X rays.

    c. ultraviolet light.

    d. infrared light.

Lesson Review

Lesson 13

Neutron Stars and Black Holes

Please Note: Use this matrix to guide your study and achieve the learning objectives of this lesson. It will also help you to view the video, which defines and demonstrates important concepts and skills as they relate to everyday life.

Learning Objective	Textbook	Telecourse Student Guide	Background Notes
1. Outline the theoretical formation of neutron stars and describe their properties.	pp. 208–209	Matching Activities: 1; Completion Activities: 1; Self-Test Questions: 1, 2.	
2. Recount the discovery of pulsars and their identification as rapidly spinning neutron stars.	pp. 209–210	Matching Activities: 2; Completion Activities: 1; Self-Test Questions: 3, 4.	
3. Describe the observed properties of pulsars and binary pulsars.	pp. 210–219	Matching Activities: 3, 4; Completion Activities: 2; Self-Test Questions: 5, 6.	
4. Describe the theoretical formation of black holes and their properties.	pp. 219–221	Matching Activities: 5, 6, 7, 8, 9, 10, 11; Completion Activities: 1, 3; Self-Test Questions: 7, 8.	
5. Recount attempts to observe black holes indirectly and list examples of black hole candidates.	pp. 221–226	Completion Activities: 1; Self-Test Questions: 9, 10.	

Unit II

Background Notes

B A C K G R O U N D   N O T E S   8

Solar Magnetic Cycle

As the sun ebbs and flows through its 11-year cycle of magnetic activity, it is not just the number of sunspots that changes. The size and shape of the corona, the intensity of the solar wind, and the incidence of solar flares also fluctuate. The high-speed electrically charged particles that escape the sun during flares are of particular concern on Earth.

In pretechnological times, the occurrence of solar flares was noticeable only in the beautiful displays of auroras (northern and southern lights) that brightened the polar skies when the solar particles plowed into Earth's magnetic field hours or days after leaving the sun. In the twenty-first century, however, the effects of solar flares have direct impact on daily life. Subtle alterations in Earth's magnetic field upset navigational equipment and induce surges in power lines. Changes in the atmosphere disrupt long-distance radio communications. Spacecraft are even more susceptible; the additional radiation near sunspot maximum heats the atmosphere so that it extends farther from Earth and increases the frictional drag on orbiting satellites. Both Skylab and the Solar Maximum

Mission experienced orbital decay, reentering and burning up in the atmosphere much sooner than would have happened if the last two sunspot maxima had not been so intense. Since both spacecraft were designed to study the sun, it is ironic that the sun was responsible for their premature demise.

On Earth, our magnetic field protects us from the brunt of these flare particles. But as humans spend more time in space beyond the protection of this field, such as when exploring the moon or Mars (which lack significant magnetism), we will have to issue flare warnings so that astronauts have time to take proper shelter from this harmful radiation. A shelter buried under a few feet of lunar or Martian soil should offer sufficient shielding.

Far from being able to predict flares, we don't even understand the 11-year cycle fully. It is a complete mystery why this cycle sometimes ceases for extended periods. The most famous instance was the Maunder minimum, first recognized by Walter Maunder about a century ago. The Maunder minimum refers to the period from 1645 to 1715, when a scarcity of sunspots, of auroral displays, and of prominent coronas during eclipses all pointed to a prolonged lack of solar activity. These years also marked the peak of the "Little Ice Age" in Europe and America, with its severe winters and short growing seasons.

Even today, scientists are uncertain whether the Maunder minimum and the Little Ice Age were related or merely coincidental. Written sunspot records go back no farther than Galileo's day, so we are unable to learn if there had been similar occurrences in the past. However, measurements of radioactive carbon 14 in tree rings can also indicate the level of solar activity during the year that each ring formed. Such studies have uncovered about a half dozen Maunder-like minima in the last 5,000 years, as well as several sustained maxima. These events correlate well with whatever temperature, winter-severity, and glacial activity data are available. Basically, when the sun is "quiet" for extended periods, Earth gets colder; when the sun is active over a sustained time, Earth get warmer.

This finding implies that the sun is actually brighter when it is covered in spots. It is difficult to measure slight changes in solar brightness from the ground because Earth's own atmosphere varies substantially in transparency over time. But recent observations from space show the sun is indeed a few tenths of a percent brighter when active. In fact, solar activity has been particularly intense during the last half century, so the global warming trend, which has so frequently been blamed on human pollution of the atmosphere with "greenhouse" gases, may actually be, at least to some degree, a by-product of natural solar variability.

Even if global warming results in part from natural phenomena, it is small consolation. In a world that already has trouble growing enough food for its burgeoning population, another mini-ice age would not be welcome. We have not been observing the sun with precision long enough to understand its true behavior; the 11-year cycles may be mere ripples upon a larger cycle. We have even been studying magnetic behavior in other stars to gain some insight into the fraction of time a star spends with its cycles either stalled or stuck in high gear. But until the results are in, the inconstancy of the sun is yet another variable to be thrown into our attempts to estimate the future of our climate and our species.

14

The Milky Way

Assignments

For the most effective study of this lesson, we suggest that you complete the assignments in the sequence listed below.

Before Viewing the Video Program

- Read the Overview and Learning Objectives for this lesson. Use the Learning Objectives to guide your reading, viewing, and thinking.

- Read Chapter 12, "The Milky Way Galaxy," pages 228–253, in the *Horizons* textbook.

- Read the Viewing Notes in this lesson.

View "The Milky Way Galaxy" Video Program

After Viewing the Video Program

- Briefly note your answers to the questions listed at the end of the Viewing Notes.

- Review all reading assignments for this lesson, especially the Chapter 12 summary on pages 249 and 252 in *Horizons* and the Viewing Notes in this lesson.

- Respond to the challenges presented by each Critical Inquiry in the reading assignment and write brief answers to the review questions at the end of Chapter 12 in *Horizons* to be certain you understand the text material.

- Complete the Review Activities in this guide to reinforce your understanding of important terms and concepts. Check your answers with the Answer Key and review when necessary.

- Take the Self-Test in this guide to measure your achievement of the Objectives. Check your answers with the Answer Key and review when necessary.

- Complete any other activities and projects assigned by your instructor.

Overview

The "astro" in astronomy means "star." But, even aside from those astronomers who have joined geologists and meteorologists in the specialized study of the nearby planets, individual stars are only about half of what most astronomers investigate. Equally important is the socializing in which stars engage as gravity binds them together into star clusters, galaxies, clusters of galaxies, and the even larger structures that were already taking shape when the universe was still very young. A complete understanding of the history of the universe, and of nature, requires that these vast stellar systems be understood as well.

This lesson concerns the galaxy of which we are a part: the Milky Way Galaxy. It is a giant, flattened, spinning, spiral-shaped assemblage of hundreds of billions of stars; our sun, with its family of planets, orbits slowly in our galaxy's outskirts. The milky glow that encircles us in very dark skies is just our inside, edge-on view of this lens-shaped galaxy.

This is a transitional lesson. It explains the role that the galaxy's spiral structure plays in triggering the formation of stars from interstellar gas clouds and shows how the heavy chemical content of succeeding generations of stars keeps increasing as a consequence of the supernova outbursts of previous generations. Thus, this lesson fills in the last remaining details of the story of stellar birth and death that has occupied us since Lesson 9.

In addition, the intimate details of our own galaxy must be compared to the billions of other galaxies we have discovered, not all of which are flattened and spiral, and not all of which are still in the business of manufacturing stars. This lesson is the first of five that look not only at galaxies and their sometimes violent interactions, but at the origin, structure, and destiny of the universe itself.

Learning Objectives

When you have completed all assignments for this lesson, you should be able to:

1. Describe the characteristics of open and globular star clusters, and explain how the study of their distribution and of variable stars and proper motion enabled Harlow Shapley to estimate the size of the Milky Way and the location of Earth's solar system within it. *HORIZONS* TEXTBOOK PAGES 230–235 (also review pages 188–189 in Chapter 10); VIDEO PROGRAM.

2. Describe the features of the Milky Way Galaxy and of its major components, including the nature of the stellar orbits in each region. *HORIZONS* TEXTBOOK PAGES 235–239; VIDEO PROGRAM.

3. Explain the importance of 21-centimeter radiation in the work of radio astronomers. *HORIZONS* TEXTBOOK PAGES 235–236, 244–246; VIDEO PROGRAM.

4. Explain the reason for differing populations of stars within the galaxy and relate their distribution to the process of the formation of the Milky Way. *HORIZONS* TEXTBOOK PAGES 239–243; VIDEO PROGRAM.

5. Describe observations of spiral arms within the Milky Way and other galaxies and explain the two major mechanisms that create spiral structure and their role in star formation. *HORIZONS* TEXTBOOK PAGES 243–248; VIDEO PROGRAM.

6. Describe observations of the nucleus of the Milky Way at nonvisible wavelengths and the nature of the suspected black hole at the center. *HORIZONS* TEXTBOOK PAGES 249–251; VIDEO PROGRAM.

Viewing Notes

After the opening relates the ancient phenomenon of the Milky Way—a narrow ribbon of fuzzy light in the night sky—to our location within the flattened disk of a huge galaxy, this video program looks at the history behind the very concept of a galaxy.

The first segment begins with William Herschel carefully counting stars in different directions to try to gauge their actual distribution in three-dimensional space. It was obvious to almost every astronomer that the stars were arranged in a flattened system, but Herschel, because of light-obscuring dust in space, incorrectly concluded that our sun was near the center of that system. The segment concentrates on the pivotal work of Harlow Shapley, who used the distribution of globular star clusters, building on the knowledge developed by Henrietta Swan Leavitt of Cepheid variable stars within them, to identify the real center and come to the all-important conclusion that it was nowhere near *us*. The galaxy was much larger than Herschel had thought and most of it is hidden from the view of ordinary telescopes by that inconvenient dust.

Armed with this new understanding of a dense, disk-shaped galaxy with a sparser, spherical halo surrounding it, the next segment considers how the Milky Way may have acquired this particular shape. Walter Baade did the pioneering work when he discovered that the population II stars in the halo, which included the globular clusters, were weak in heavy elements or "metals," but that the population I stars in the disk, including the newly forming young star clusters in the spiral arms, were much richer in these elements.

Because the metal content of the interstellar medium is continuously enriched by the explosive deaths of old, massive stars, and because the halo only contains stars with very lengthy life spans, it became clear that population II stars are older than population I stars and that the halo must have formed earlier than the disk! Thus the segment finishes with the traditional explanation for the formation of our galaxy from a huge spherical cloud that collapsed into a smaller, flatter, faster-spinning disk. But you will see in future programs that this simple scenario was probably modified by the collisions and mergers of smaller gas clouds and/or galaxies at a number of moments during our galaxy's history.

The final segment explores the details of our galaxy's structure, beginning with the use of 21-centimeter radio observations to trace out the location of interstellar hydrogen, and the spiral arms themselves, over most of the galaxy's disk. It then looks at density waves, one of the two processes that can create and maintain a spiral pattern (the other, self-sustaining star formation, is described in the text). These density waves are responsible for initiating the waves of hot, bright, blue young stars and open star clusters which make spiral arms stand out so noticeably in photographs of other spiral galaxies. Finally, the program turns to the massive, dark and mysterious black hole that may be buried at the galaxy's very center. This, too, is just a taste of the strange behavior you'll see in later programs about the other galaxies beyond ours.

As you watch the video program, consider the following questions:

1. Why do we see the Milky Way encircling us at night?

2. Why did Herschel conclude that the Milky Way was flattened and our solar system near the center?

3. Why did Shapley conclude that our solar system was not near the center of the Milky Way?

4. What are the major regions, or parts, of our galaxy: the Milky Way? What are the relative ages of stars in each region? How do we know? What does this tell us about the formation of the Milky Way?

5. How do we study the spiral structure of the Milky Way when interstellar dust obscures most of it?

6. What process can create and maintain a spiral pattern, and how does this process relate to star formation?

7. At what wavelengths can we "see" the center of the Milky Way, and what do we find there?

Review Activities

Matching

Match the terms listed below with the definitions that follow. Check your answers with the Answer Key and review any terms you missed.

_____ 1. Andromeda Galaxy

_____ 2. instability strip

_____ 3. period of a variable star

_____ 4. RR Lyrae variable star

_____ 5. Cepheid variable star

_____ 6. period-luminosity relation

_____ 7. open star cluster

_____ 8. globular star cluster

_____ 9. proper motion

_____ 10. calibration

_____ 11. disk component

_____ 12. kiloparsec (kpc)

_____ 13. spiral arm

_____ 14. spherical component

_____ 15. halo

_____ 16. nuclear bulge

_____ 17. differential rotation

_____ 18. rotation curve

_____ 19. galactic corona

_____ 20. population I and II

_____ 21. metal

_____ 22. spiral tracer

_____ 23. density wave theory

_____ 24. flocculent galaxy

_____ 25. self-sustaining star formation

_____ 26. Sagittarius A *

a. the part of the galaxy including the halo and nuclear bulge

b. the part of the galaxy confined to its plane of rotation

c. the spherical cloud of stars at the center of spiral galaxies

d. large spherical region of a galaxy containing globular clusters

e. region of disk containing bright stars, star clusters, gas, and dust

f. an object, such as an O or a B star, used to map spiral arms

g. hypothesized mass outside the halo

h. small group of relatively young stars in the disk

i. large group of old stars in the halo and nuclear bulge

j. visible to our unaided eyes as a faint patch of light, it probably looks much like, but is not part of, the Milky Way Galaxy

k. region of H-R diagram in which stars pulsate in diameter

l. length of time it takes a variable star to pulsate once

m. variation in the length of the orbital periods of different stars

n. graph of orbital velocity versus radius in the disk of a galaxy

o. intense radio source near the center of the Milky Way

p. explanation of spiral arms as compressions of the interstellar medium

q. explanation of spiral arms as stellar birth triggered by previous stellar birth

r. a type of galaxy with many short spiral segments

s. a comparison of pulsation time and intrinsic brightness

t. establishment of the relationship between two parameters

u. low-mass star that regularly pulsates in less than one day

v. high-mass star that takes more than one day to pulsate regularly

w. rate at which a star moves across the sky

x. a little over 3,000 light-years

y. an atom heavier than helium

z. a stellar distinction based on the abundance of metal

Completion

Fill each blank with the most appropriate term from the list for that paragraph. A term may be used once, more than once, or not at all. Check your answers with the Answer Key and review when necessary.

1. Regarding the age of stars, blue stars tend to be _____ than red stars; the stars in a region containing only red main-sequence stars and giant stars are probably _____. Globular clusters contain mostly_____ stars and open clusters contain relatively _____ stars. In a galaxy, the halo and nuclear bulge contain mostly _____ stars, and the disk and spiral arms contain mostly _____ stars. Population I stars are _____ and population II stars are _____. A star that is very metal-poor is _____.

   older   younger

2. A spiral galaxy with just two well-defined arms is known as a
_____ galaxy and probably formed as a result of
_____ compressing the interstellar material to trigger star
formation. A spiral galaxy with numerous arms, branches, and spurs
is known as a _____ galaxy and probably resulted from
_____ regions being wound up by _____.

density waves   grand design

differential rotation   self-sustaining star
                                                            formation
flocculent

Self-Test

Select the one best answer for each question. Check your answers with the
Answer Key and review when necessary.

1. Compared to open star clusters, globular clusters tend to be

   a. richer in metals and younger.

   b. poorer in metals and younger.

   c. richer in metals and older.

   d. poorer in metals and older.

2. A group of 100 to 1,000 stars in a region 3 to 30 parsecs in diameter
   would be

   a. a galaxy.

   b. an open cluster.

   c. an association.

   d. a globular cluster.

3. Of the following stars, the one that has the longest period of variability is

   a. an RR Lyrae star.

   b. the least luminous Cepheid.

   c. the most luminous Cepheid.

   d. a magnetar.

4. The approximate diameter of the disk of the Milky Way Galaxy is
   a. 1 light-year.

   b. 100 light-years.

   c. 75,000 light-years.

   d. 10 million light-years.

5. In a typical spiral galaxy, stars in the
   a. disk have nearly circular orbits.

   b. disk have highly elliptical orbits.

   c. halo have nearly circular orbits.

   d. galactic plane have highly elliptical orbits.

6. As you go outward from the sun toward the edge of the Milky Way's disk, the orbital velocity of the stars
   a. decreases greatly (Keplerian motion).

   b. decreases slightly.

   c. remains exactly the same.

   d. increases slightly.

7. The 21-centimeter line of hydrogen is in the
   a. gamma ray part of the spectrum.

   b. ultraviolet part of the spectrum.

   c. infrared part of the spectrum.

   d. radio part of the spectrum.

8. Because radio astronomers can detect the 21-centimeter wavelength of _____ in space, they can use this radiation to map the galaxy.
   a. helium produced by nuclear fusion

   b. neutral (un-ionized) hydrogen

   c. lithium

   d. X rays

9. Which of the following lists of galactic regions is correctly listed in order of increasing metal content of stars commonly found in each region?

    a. disk, halo, spiral arms, nuclear bulge

    b. halo, spiral arms, disk, nuclear bulge

    c. spiral arms, halo, nuclear bulge, disk

    d. halo, nuclear bulge, disk, spiral arms

10. In terms of the observable metal content in the outer layers of stars,

    a. older stars are metal-poor because their nuclear reactions have consumed the metals used in their formation.

    b. older stars are metal-rich because their nuclear reactions formed increasingly heavy elements over time.

    c. younger stars are metal-rich because they formed from gases enriched by the heavy elements created in the cores of previous stars.

    d. younger stars are metal-poor because it takes much time for heavy elements to condense out of the interstellar gas.

11. In order to trace the spiral structure of the Milky Way over most of its disk, we need to make observations at

    a. gamma ray wavelengths.

    b. X-ray wavelengths.

    c. infrared wavelengths.

    d. radio wavelengths.

12. The number of arms in the Milky Way is

    a. one.

    b. exactly two.

    c. at least two, but with branches and spurs.

    d. none, only a great many short spiral segments.

13. The galactic center of the Milky Way lies in the constellation

    a. Cygnus.

    b. Orion.

    c. Sagittarius.

    d. Perseus.

14. In terms of mass, the suspected black hole at the center of the Milky Way is

    a. 30 solar masses.

    b. 3,000 solar masses.

    c. 3 million solar masses.

    d. 300 billion solar masses.

Lesson Review

Lesson 14

The Milky Way

Please Note: Use this matrix to guide your study and achieve the learning objectives of this lesson. It will also help you to view the video, which defines and demonstrates important concepts and skills as they relate to everyday life.

Learning Objective	Textbook	Telecourse Student Guide	Background Notes
1. Describe the characteristics of open and globular star clusters, and explain how the study of their distribution and of variable stars and proper motion enabled Harlow Shapley to estimate the size of the Milky Way and the location of Earth's solar system within it.	pp. 188–189, 230–235	Matching Activities: 1, 2, 3, 4, 5, 6, 7, 8, 9, 10; Self-Test Questions: 1, 2, 3, 4.	
2. Describe the features of the Milky Way Galaxy and of its major components, including the nature of the stellar orbits in each region.	pp. 235–239	Matching Activities: 12, 13, 14, 15, 16, 17, 18, 19; Self-Test Questions: 5, 6.	
3. Explain the importance of 21-centimeter radiation in the work of radio astronomers.	pp. 235–236, 244–246	Self-Test Questions: 7, 8.	
4. Explain the reason for differing populations of stars within the galaxy and relate their distribution to the process of the formation of the Milky Way.	pp. 239–243	Matching Activities: 20, 21; Completion Activities: 1; Self-Test Questions: 9, 10.	
5. Describe observations of spiral arms within the Milky Way and other galaxies and explain the two major mechanisms that create spiral structure and their role in star formation.	pp. 243–248	Matching Activities: 22, 23, 24, 25; Completion Activities: 2; Self-Test Questions: 11, 12.	

Learning Objective	Textbook	Telecourse Student Guide	Background Notes
6. Describe observations of the nucleus of the Milky Way at nonvisible wavelengths and the nature of the suspected black hole at the center.	pp. 249–251	Matching Activities: 26; Self-Test Questions: 13, 14.	

15

Galaxies

Assignments

For the most effective study of this lesson, we suggest that you complete the assignments in the sequence listed below.

Before Viewing the Video Program

- Read the Overview and Learning Objectives for this lesson. Use the Learning Objectives to guide your reading, viewing, and thinking.

- Read Chapter 13, "Galaxies," pages 254–275, in the *Horizons* textbook.

- Read the Viewing Notes in this lesson.

View the "Galaxies" Video Program

After Viewing the Video Program

- Briefly note your answers to questions listed at the end of the Viewing Notes.

- Review all reading assignments for this lesson, especially the Chapter 13 summary on page 274 in *Horizons* and the Viewing Notes in this lesson.

- Respond to the challenges presented by each Critical Inquiry in the reading assignment and write brief answers to the review questions at the end of Chapter 13 in *Horizons* to be certain you understand the text material.

- Complete the Review Activities in this guide to reinforce your understanding of important terms and concepts. Check your answers with the Answer Key and review when necessary.

- Take the Self-Test in this guide to measure your achievement of the Objectives. Check your answers with the Answer Key and review when necessary.

- Complete any other activities and projects assigned by your instructor.

Overview

It was not until the second decade of the twentieth century that astronomers began to develop a reasonably correct understanding of the true size and shape of the Milky Way Galaxy and of our solar system's location within it. But many questions about our galaxy remained. For example, was our galaxy the entire universe, or were other galaxies sprinkled throughout an even vaster space? Although astronomers had identified objects they called "spiral nebulae," they did not agree on what these objects were. Some astronomers thought these nebulae were other galaxies, while others thought they were merely spiral clouds of gas within the boundaries of the Milky Way.

In 1920, a debate between Harlow Shapley and Heber Curtis centered on one such object: the Great Spiral Nebula in Andromeda, also known as M 31. As the largest and brightest of these objects, it was presumably among the closest. Shapley thought it was a cloud of gas within the Milky Way; Curtis thought it was a galaxy outside the Milky Way. The issue was resolved by 1924, when Edwin Hubble turned the 100-inch Mount Wilson telescope, then the world's largest, in the direction of M 31. He saw that it was composed of a huge number of extremely faint stars. The Great Spiral Nebula was indeed a galaxy, as were all the other spiral nebulae at even greater distances.

With Hubble's discovery, the universe had suddenly become a much larger place, and astronomers had an entirely new class of phenomena to study and understand. Hubble began by classifying galaxies according to their shapes; although many were spiral like the Milky Way Galaxy, others were elliptical or irregular. Hubble and other astronomers also measured the distance, size, brightness, and mass of individual galaxies, much as astronomers had started to do for stars a century earlier.

But unlike the stars of the Milky Way, which both approach and recede from us as they race our sun around the galaxy's center, the distant galaxies were found to move in a single direction: away from us. Hubble also discovered a simple relationship—the farther a galaxy is from us, the faster it is moving away from us. This Hubble law was the basis of the conclusion that the universe is expanding, possibly as a result of a "big bang" about 14 billion years ago. Although this lesson describes the formation of galaxies in the early universe and the manner in which they are distributed throughout space, Lessons 17 and 18 provide a more complete description of how the distribution originated as well as the many other implications of a big bang.

Learning Objectives

When you have completed all assignments for this lesson, you should be able to:

1. Identify the three major morphological classes of galaxies and describe their characteristics. *HORIZONS* TEXTBOOK PAGES 255–259; VIDEO PROGRAM.

2. Describe how distance, diameter, luminosity, and mass of a galaxy are measured, and show how these imply the existence of dark matter. *HORIZONS* TEXTBOOK PAGES 259–267.

3. State the Hubble law and explain the result of Hubble's comparison of the distances and radial velocities of distant galaxies. *HORIZONS* TEXTBOOK PAGES 262–263.

4. Describe the factors involved in galactic formation, including the role of collisions between galaxies. *HORIZONS* TEXTBOOK PAGES 267–274; VIDEO PROGRAM.

Viewing Notes

The first segment of the video program "Galaxies" concerns one of the great ironies of modern astronomy. Harlow Shapley had just used Cepheid variable stars to measure the distances to globular clusters and thus pinpoint the center of our galaxy. Yet, he was one of those astronomers who erroneously concluded that our galaxy was alone in the universe. Just a few years later, Edwin Hubble used the same observatory as Shapley to spot Cepheids in the Andromeda Galaxy and prove it was a huge system of stars, a galaxy in its own right, far beyond the edges of our own. We now know there are as many other galaxies spread throughout space as there are stars within the Milky Way, which is a lot! And if the vastness of the universe were not enough of a shock, Hubble's comparison of galactic distances and radial velocities revealed that the universe is becoming even vaster as it expands with time. These discoveries, as described in the video program by Sallie Baliunas, an astronomer at the Smithsonian Astrophysical Observatory, took "the last great step in the Copernican revolution, displacing Earth from the center of the universe, the sun from the center of the universe, and now the galaxy from the center of the universe."

As the next segment explains, all galaxies are not just like ours; they come in a variety of sizes and shapes, which Hubble himself thought might represent galaxies in different stages of their lives. But later observations showed that this was not the case. Elliptical galaxies were observed to contain only old stars and little gas and dust, because they had efficiently converted all their gas into stars early in their existence. Spiral galaxies, on the other hand, contained similarly old stars in their halos and central bulges, but contained much younger stars, along with much gas and dust, in their disks and arms; their highly flattened disks preserved some of their original gas, allowing star formation to continue there to the present day.

But there was no type of nearby galaxy that contained only young stars. As the expanding universe cooled and thinned, there was an era more than 10 billion years ago when the conditions were appropriate for galaxies to form, so they did. But the conditions have changed since then, and the universe isn't making galaxies any more. This segment also looks at galactic formation. Although the various shapes seem to be the result of galaxies forming with different circumstances, such as rate of spin, it has become increasingly apparent that collisions and mergers between galaxies, mostly early in their existence, played an even larger role in defining their final form. Indeed, no galaxy's form can be considered final as long as the opportunity for a future collision exists.

But how can galaxies in an expanding universe collide? The final segment answers this question by revealing that galaxies exist within clusters, such as the Milky Way's own Local Group, in which the galaxies remain gravitationally bound together despite the expansion. Our galaxy was recently discovered to be swallowing a small companion approaching from the other side of the Milky Way's center, and it looks as though the Magellanic clouds and even the Andromeda Galaxy may be in for a merger with us at some point in the future. But gravity doesn't stop there. It binds clusters into superclusters, and superclusters into vast walls and filaments of galaxies that are separated by vast voids. The large-scale structure of the universe is frothy. And an effect of this lumpiness is to cause galaxies to stream through space with peculiar motions that are more complicated than that of a simple expansion. Although the program hints at the origin of this structure, it is Lessons 17 and 18 that reveal how the seeds of the galaxies may have been planted during the Big Bang itself and nurtured by the gravity of a mysterious dark matter that has yet to be identified.

As you watch the video program, consider the following questions:

1. Why was there disagreement over whether other galaxies existed beyond ours, and how was that controversy resolved?

2. From our knowledge of a galaxy's radial velocity (the speed with which it approaches or recedes from us), how can we estimate its distance?

3. What are the characteristics of the different types of galaxies?

4. How do we think galaxies of different shapes formed, and what role did collisions play in that process?

5. What is the significance of the fact that the oldest stars in any type of galaxy are about the same age? What does that say about an evolving universe?

6. How are galaxies distributed throughout the universe? How does this distribution affect the motions of galaxies?

Review Activities

Matching

Match the terms listed below with the definitions that follow. Check your answers with the Answer Key and review any terms you missed.

_____ 1. elliptical galaxy

_____ 2. spiral galaxy

_____ 3. barred spiral galaxy

_____ 4. irregular galaxy

_____ 5. megaparsec (Mpc)

_____ 6. distance indicator

_____ 7. H II region

_____ 8. look-back time

_____ 9. Hubble law

_____ 10. Hubble constant

_____ 11. cluster method

_____ 12. velocity dispersion method

_____ 13. dark matter

_____ 14. rich galaxy cluster

_____ 15. poor galaxy cluster

_____ 16. ring galaxy

_____ 17. galactic cannibalism

_____ 18. Magellanic clouds

a. the direct linear proportion between galactic distances and red shifts

b. a number that indicates how fast a galaxy at any given distance is moving away from us

c. an object of known luminosity or diameter in another galaxy

d. material so far detectable only by its gravitational influence

e. amount by which one sees into the past by observing an object at a given distance

f. small galaxies beyond, but close to, the Milky Way Galaxy, which appear to have passed through the outer disk of our galaxy 200 million years ago and may eventually merge with it

g. determination of a galaxy's mass using the range of velocities within a galaxy

h. determination of a galaxy's mass based on the motions of galaxies within a cluster

i. a little over 3 million light-years

j. cloud of ionized hydrogen that can be used as a distance indicator

k. believed to result from a head-on collision of two galaxies

l. process by which a large galaxy absorbs a smaller one

m. irregularly shaped group of galaxies, many of them spiral, with no giant ellipticals

n. group of mostly elliptical galaxies spread over a spherical volume about 3 Mpc wide

o. armless, chaotically shaped galaxy with lots of gas and dust, and both young and old stars

p. a round or slightly flattened galaxy with mostly old stars and little gas and dust

q. a highly flattened galaxy with a well-defined disk containing gas, dust, and young stars arranged in curved arcs

r. a galaxy with an elongated nucleus

Completion

Fill each blank with the most appropriate term from the list for that paragraph. A term may be used once, more than once, or not at all. Check your answers with the Answer Key and review when necessary. If a question requires two or more answers in succession, they may be in any order, unless the question indicates otherwise.

1. The type of galaxy with the least dust and gas, and fewest hot stars, is the _____ galaxy. These galaxies are classified into subgroups, ranging from _____ to _____. A galaxy with a disk and lots of gas, dust, and star formation is of the _____ type. A disk galaxy with no arms, little gas and dust, and only a few hot, bright stars is known as a(n) _____ galaxy. This type is thought to be intermediate between _____ and _____ galaxies. A galaxy with much gas, dust, and star formation but no disk or nucleus is classified as _____. Most of the galaxies that have been catalogued are _____, but the most common kind of galaxy is believed to be _____. A rich galaxy cluster would contain almost all _____ and _____ galaxies.

E0	Sa
E7	Sc
elliptical	S0
irregular	spiral

2. Observing an object of known diameter or luminosity in another galaxy can tell us that galaxy's _____. This quantity can also be measured by knowing the Hubble constant and the galaxy's _____. From this information and apparent brightness, we can determine the object's _____. From this information and angular size, we can determine the object's _____. Observations of the velocities of stars within galaxies or of galaxies within clusters can yield information on a galaxy's _____. The conclusion that the vast bulk of matter in most galaxies is dark was reached through comparisons of the _____ and _____ of galaxies.

diameter	mass
distance	radial velocity
luminosity	

3. Once astronomers were able to identify and map distant galaxies, they soon realized that galaxies are not distributed _____. Rather, galaxies are grouped into _____. A group of galaxies with more than 1,000 galaxies is known as a _____. Most of the galaxies within this group are _____, and often have one or more _____ galaxies at their centers. Our Milky Way is part of a grouping known as the _____, which is an example of a _____. Such a group contains fewer than 1,000 galaxies and its shape is _____.

clusters	Local Zone
dwarf	poor galaxy cluster
elliptical	rich galaxy cluster
giant	spiral
irregular	uniformly
Local Group	

Self-Test

Select the one best answer for each question. Check your answers with the Answer Key and review when necessary.

1. Of the following types of galaxies, the one that is most circular in apparent shape and featureless is
   a. E0.
   b. E7.
   c. S0.
   d. Sb.

2. Of the following types of galaxies, the one that has the most gas and dust is
   a. E6.
   b. S0.
   c. Sa.
   d. SBc.

3. Which type of galaxy has the greatest range in diameter, mass, and luminosity?
   a. a spiral galaxy
   b. an elliptical galaxy
   c. an irregular galaxy
   d. All of the above have similar ranges.

4. For nearer galaxies, the most accurate distance indicators are
   a. H II regions.
   b. supernovae.
   c. planetary nebulae.
   d. Cepheid variables.

5. When astronomers measure the masses of galaxies and find that the measured masses are much too large, the cause is probably invisible material called

   a. dark matter.

   b. black holes.

   c. brown dwarfs.

   d. neutrinos.

6. If the Hubble constant equals 50 km/sec/Mpc, and a galaxy has a recessional velocity of 5,000 km/sec, then its distance would be

   a. 2.5 Mpc.

   b. 50 Mpc.

   c. 100 Mpc.

   d. 5000 Mpc.

7. Excluding the galaxies in the Local Group, the other galaxies show spectra that are

   a. all blue shifted.

   b. all red shifted.

   c. about equally divided between blue shifted and red shifted.

   d. not shifted at all.

8. Collisions between galaxies

   a. are extremely rare.

   b. occur once in a while, but the stars within them almost never collide.

   c. are frequent, causing most of the stars within them to collide also.

   d. are frequent, but have little effect on either galaxy.

9. When the stars of a small galaxy are spread through a larger one, this merging is

   a. the creation of a cluster.

   b. a local grouping.

   c. a supernova.

   d. called galactic cannibalism.

10. Based on our current understanding of galactic formation and evolution, and look-back time,

    a. nearby rich galaxy clusters consist mainly of elliptical galaxies, although more distant younger clusters should have a somewhat larger number of spiral galaxies.

    b. nearby rich galaxy clusters consist mainly of spiral galaxies, although more distant older clusters should have a somewhat larger number of elliptical galaxies.

    c. rich galaxy clusters should contain the same percentage of elliptical galaxies regardless of distance.

    d. rich galaxy clusters should contain the same percentage of elliptical galaxies as poor galaxy clusters.

Lesson Review

Lesson 15

Galaxies

Please Note: Use this matrix to guide your study and achieve the learning objectives of this lesson. It will also help you to view the video, which defines and demonstrates important concepts and skills as they relate to everyday life.

Learning Objective	Textbook	Telecourse Student Guide	Background Notes
1. Identify the three major morphological classes of galaxies and describe their characteristics.	pp. 255–259	Matching Activities: 1, 2, 3, 4, 18; Completion Activities: 1; Self-Test Questions: 1, 2.	
2. Describe how distance, diameter, luminosity, and mass of a galaxy are measured, and show how these imply the existence of dark matter.	pp. 259–267	Matching Activities: 5, 6, 7, 8, 11, 12, 13, ; Completion Activities: 2; Self-Test Questions: 3, 4, 5.	
3. State the Hubble law and explain the result of Hubble's comparison of the distances and radial velocities of distant galaxies.	pp. 262–263	Matching Activities: 9, 10; Completion Activities: 2; Self-Test Questions: 6, 7.	
4. Describe the factors involved in galactic formation, including the role of collisions between galaxies.	pp. 267–274	Matching Activities: 14, 15, 16, 17; Completion Activities: 3; Self-Test Questions: 8, 9, 10.	

LESSON

16

Peculiar Galaxies

Assignments

For the most effective study of this lesson, we suggest that you complete the assignments in the sequence listed below.

Before Viewing the Video Program

- Read the Overview and Learning Objectives for this lesson. Use the Learning Objectives to guide your reading, viewing, and thinking.

- Read Chapter 14, "Galaxies with Active Nuclei," pages 276–294, in the *Horizons* textbook.

- Read the Viewing Notes in this lesson.

View the "Peculiar Galaxies" Video Program

After Viewing the Video Program

- Briefly note your answers to questions listed at the end of the Viewing Notes.

- Review all reading assignments for this lesson, especially the Chapter 14 summary on page 293 in *Horizons* and the Viewing Notes in this lesson.

- Respond to the challenges presented by each Critical Inquiry in the reading assignment and write brief answers to the review questions at the end of Chapter 14 in *Horizons* to be certain you understand the text material.

- Complete the Review Activities in this guide to reinforce your understanding of important terms and concepts. Check your answers with the Answer Key and review when necessary.

- Take the Self-Test in this guide to measure your achievement of the Objectives. Check your answers with the Answer Key and review when necessary.

- Complete any other activities and projects assigned by your instructor.

Overview

The previous lesson revealed that the universe is an enormous place, populated by billions of galaxies with billions of stars apiece. But it may also have left the impression that a galaxy is little more than a collection of stars, bound together forever by their mutual gravity and shining in unison.

Although stars may represent the bulk of a galaxy's visible matter, it became clear that some galaxies have something else very strange going on within them, apparently deep down in their centers. This lesson explores these "peculiar," or active, galaxies. Some are so bright, compact, and distant that they once challenged our most basic conclusions about space and time. Active galaxies, which produce far more energy over many different wavelengths than ordinary starlight can account for, seemed to be more and more commonplace at greater distances, that is, when we looked farther into the past toward an ever younger universe.

The first active galaxies, discovered in the 1950s, were radio galaxies. These galaxies release radio energy from two large lobes, or areas of the sky, extending from both sides of the visible galaxy and going far beyond the visible galaxy's borders. Later, in the early 1960s, quasars were discovered. Quasars are so extremely distant that they appear as little more than starlike points of light, but have some of the most unusual properties of all active galaxies.

One of the greatest accomplishments of recent research was the realization that a single phenomenon, the presence of a single, supermassive black hole at the centers of some galaxies, can explain quasars as well as the complete variety of other active galaxies. And the steady decline in galactic nuclear activity as we go from the distant (and young) quasars to nearer active galaxies and, finally, to the nearly normal, closest (and oldest) galaxies, including our own Milky Way, may simply be the result of a steady reduction in the amount of material available to fall into these black holes and trigger their tremendous energy output. Indeed, as both galaxies and the universe age, they all appear to become more mellow. This is exactly the kind of evolving universe predicted by the Big Bang theory, which will be the focus of the next two lessons.

Learning Objectives

When you have completed all assignments for this lesson, you should be able to:

1. Describe observations of active galaxies and the mechanisms proposed to explain their energy output and other characteristics. *HORIZONS* TEXTBOOK PAGES 277–285; VIDEO PROGRAM.

2. Describe the discovery and characteristics of quasars. *HORIZONS* TEXTBOOK PAGES 285–288; VIDEO PROGRAM.

3. Explain how quasar distances are estimated and how recent observations contradict some earlier explanations of quasars. *HORIZONS* TEXTBOOK PAGES 288–290; VIDEO PROGRAM.

4. Describe the current explanation of quasars and their energy sources and the implications of this explanation for the formation of galaxies and early history of the universe. *HORIZONS* TEXTBOOK PAGES 290–292; VIDEO PROGRAM.

Viewing Notes

The program "Galaxies with Active Nuclei" begins with what was originally the greatest mystery—the quasars. When first discovered in 1960, they had an almost starlike appearance but very non-starlike spectra. The identification of very large red shifts in those spectra a few years later suggested that quasars were also very distant, residing among and beyond the farthest galaxies in the universe. But these huge distances also implied that quasars were as luminous as about a thousand of our Milky Way galaxies rolled into one.

That might not have been a problem if quasars could have been unusually large galaxies with correspondingly enormous numbers of stars. But their tiny appearance, coupled with extremely rapid variations in their brightness, showed that all of that energy had to be coming from a volume of space that was minuscule even when compared to just one normal galaxy. And so the quasar controversy was born: how does so much energy arise from so little space?

The second segment, "Shocking Red Shifts," shows that this question may have been answered. Astronomers have come to realize that a supermassive black hole, with the equivalent mass of millions to billions of stars, is likely to form in the very centers of the larger galaxies. Although no energy can be emitted by a black hole, material falling into a black hole can be accelerated and heated and give off tremendous energy (remember the accretion disk around the "ordinary" black hole in Cygnus X-1, described in Lesson 13).

Because of their great distances, we are seeing quasars at a very early moment in the universe's history, when the galaxies were first forming. At that early time, there was more "food" available to fall into these holes and create released energy. The galaxies had more gas in them then because not as much of the gas had been converted into stars yet. And collisions and mergers between galaxies were more frequent, providing perhaps the most efficient way of bringing fresh gas into the vicinity of a black hole to feed it. Once formed, these black holes never go away. They would simply become increasingly less active as less food was available for them. For that reason, it is not surprising that some of the larger nearby galaxies, including the Milky Way, show evidence of relatively inactive black holes in their nuclei.

The final segment extends these concepts to explain the zoo of different kinds of active galaxies, such as Seyferts and radio galaxies, which lie at intermediate distances and, thus, are seen at ages that are intermediate between the young quasars and older, nearby, normal galaxies. With lesser gas content and less frequent collisions, these active galaxies show similar but less intense activity than the quasars. But we have learned that it is not just the amount of gas feeding a hole that determines what we see, but also the angle at which we view the galaxy!

As the environment surrounding a supermassive black hole is described in more detail, we see that the inner, hotter part of the accretion disk produces most of the energy, but there is a larger, outer torus (donut-shaped region) present as well which can be laden with light-obscuring dust. Seen from the edge, this dust torus can prohibit a view of the smaller, brighter core inside it, although the program does provide an example of how clever observations can sometimes reveal the core to us anyway. And two, rapid beams of material ejected perpendicular to the torus can explain the jets and double-lobed radio emission observed around some quasars and active galaxies as well.

As you watch the video program, consider the following questions:

1. What reasons led astronomers to conclude that quasars are not stars?

2. Why have astronomers concluded that quasars are extremely distant, luminous, small, and young?

3. How does a supermassive black hole explain the properties of quasars?

4. How does a supermassive black hole explain the properties of active galaxies?

5. How does the angle at which we view an active galaxy affect what we see?

6. Why does the intensity of galactic activity change as we look to greater distances?

Review Activities

Matching

Match the terms listed below with the definitions that follow. Check your answers with the Answer Key and review any terms you missed.

_____ 1. radio galaxy

_____ 2. active galaxy

_____ 3. AGN

_____ 4. double-lobed

_____ 5. hot spot

_____ 6. double-exhaust model

_____ 7. Seyfert galaxy

_____ 8. BL Lac object

_____ 9. quasar

_____ 10. relativistic red shift

_____ 11. gravitational lens

a. small, powerful source of energy believed to be the active core of a very distant galaxy

b. the change in wavelength in the light from an object receding at a speed near that of light

c. a spiral galaxy with an unusually bright, small, active core

d. center of active galaxy that is emitting large amounts of excess energy

e. a galaxy emitting substantial unusual radiation, sometimes in the form of radio waves, which cannot be explained by ordinary starlight

f. active core of an elliptical galaxy

g. a galaxy emitting substantial energy at radio wavelengths

h. a galaxy in which radio waves are emitted from two large regions on opposite sides of the galaxy

i. the proposed explanation for a double-lobed galaxy

j. an area of especially intense emission in a double-lobed galaxy

k. an effect caused by a nearer galaxy's gravity on the image of a quasar

Completion

Fill each blank with the most appropriate term from the list for that paragraph. A term may be used once, more than once, or not at all. Check your answers with the Answer Key and review when necessary.

1. The spectral lines of quasars tend to have very _____ shifts, and these shifts are always toward the _____ end of the spectrum. From this shift, astronomers have concluded that quasars are very _____, that their luminosity is very _____, and that their energy comes from very _____ regions, because their energy varies over very _____ intervals of time. At a quasar's center, astronomers theorize that there is a _____, and its mass is very _____. In comparison to other galaxies, quasars are thought to have formed much _____ and quasars are _____ likely to interact with other galaxies.

black hole	low
blue	more
distant	near
earlier	nucleus
high	red
large	small
less	

Self-Test

Select the one best answer for each question. Check your answers with the Answer Key and review when necessary.

1. An eruption in the nucleus of an active galaxy can likely be triggered by
   a. a chain reaction of supernovae.
   b. a collision or tidal interaction with other galaxies.
   c. a big bang.
   d. the collapse of an iron core.

2. The radio emission from radio galaxies typically comes
   a. exclusively from the tiny core of each galaxy.
   b. from the same regions that emit visible light.
   c. from two large regions beyond and on opposite sides of the galaxy.
   d. uniformly from a spherical halo surrounding the galaxy.

3. Of the following objects, the one that **MUST** be a spiral galaxy is a
   a. quasar.
   b. radio galaxy.
   c. BL Lac object.
   d. Seyfert.

4. Most quasars have all of the following characteristics **EXCEPT**
   a. a strong radio source.
   b. a large red shift.
   c. a starlike appearance.
   d. superluminosity.

5. Most of the lines in the spectrum of the quasar 3C 273 are caused by
   a. calcium.
   b. carbon.
   c. helium.
   d. hydrogen.

6. The speed of a very highly redshifted quasar as determined by the relativistic Doppler formula is

    a. greater than that determined by the classical formula.

    b. exactly the same as that determined by the classical formula.

    c. less than that determined by the classical formula.

    d. almost impossible to determine because of how far quasars are from Earth.

7. Quasar distances are estimated by

    a. their observed brightness.

    b. observations of Cepheids within them.

    c. parallax.

    d. their red shifts and the Hubble law.

8. All of the following statements are evidence that quasars are extremely distant **EXCEPT** for the

    a. observation of a supernova in a quasar's host galaxy.

    b. apparent connection of some quasars to galaxies of lesser red shift.

    c. detection of fuzzy light of similar red shift around some quasars.

    d. gravitational lensing of some quasars by nearer galaxies.

9. Quasars are thought to be galaxies

    a. early in their lives, although they formed early in the universe.

    b. early in their lives, having formed recently.

    c. at about the same stage of their lives as the Milky Way.

    d. in their final stages of existence, having formed long ago.

10. Of the following objects, the one that produces the **LEAST** energy is a

    a. supernova.

    b. Seyfert galaxy.

    c. BL Lac object.

    d. quasar.

Lesson Review

Lesson 16

Peculiar Galaxies

Please Note: Use this matrix to guide your study and achieve the learning objectives of this lesson. It will also help you to view the video, which defines and demonstrates important concepts and skills as they relate to everyday life.

Learning Objective	Textbook	Telecourse Student Guide	Background Notes
1. Describe observations of active galaxies and the mechanisms proposed to explain their energy output and other characteristics.	pp. 277–285	Matching Activities: 1, 2, 3, 4, 5, 5, 7, 8; Self-Test Questions: 1, 2, 3.	
2. Describe the discovery and characteristics of quasars.	pp. 285–288	Matching Activities: 9; Completion Activities: 1; Self-Test Questions: 4, 5.	
3. Explain how quasar distances are estimated and how recent observations contradict some earlier explanations of quasars.	pp. 288–290	Matching Activities: 10, 11; Self-Test Questions: 6, 7, 8.	
4. Describe the current explanation of quasars and their energy sources and the implications of this explanation for the formation of galaxies and early history of the universe.	pp. 290–292	Completion Activities: 1; Self-Test Questions: 9, 10.	

LESSON
17

The Big Bang

Assignments

For the most effective study of this lesson, we suggest that you complete the assignments in the sequence listed below.

Before Viewing the Video Program

- Read the Overview and Learning Objectives for this lesson. Use the Learning Objectives to guide your reading, viewing, and thinking.

- Read Chapter 15, "Cosmology in the 21st Century," pages 295–321, in the *Horizons* textbook.

- Read the Viewing Notes in this lesson.

View "The Big Bang" Video Program and Review the final segment of Video Program 15, "Galaxies"

After Viewing the Video Program

- Briefly note your answers to questions listed at the end of the Viewing Notes.

- Review all reading assignments for this lesson, especially the Chapter 15 summary on pages 319–320 in *Horizons* and the Viewing Notes in this lesson.

- Respond to the challenges presented by each Critical Inquiry in the reading assignment and write brief answers to review questions 1–8 at the end of Chapter 15 in *Horizons* to be certain you understand the text material.

- Complete the Review Activities in this guide to reinforce your understanding of important terms and concepts. Check your answers with the Answer Key and review when necessary.

- Take the Self-Test in this guide to measure your achievement of the Objectives. Check your answers with the Answer Key and review when necessary.

- Complete any other activities and projects assigned by your instructor.

Overview

As the lessons of this telecourse have been drawing us ever farther from Earth and deeper into the past, we have finally arrived at our ultimate destination. This and the next lesson find us poised at the edge of the observable universe and at the very beginning of time. The subject for these lessons is cosmology, the branch of astronomy that is concerned with the universe as a whole, without regard for the little things like stars and planets and people. Instead, cosmologists pursue answers to the grandest questions about the origin, structure, and fate of the cosmos. It is the *big* picture in the truest sense of that phrase.

This is such a vast subject, and such an active area of research, that it was given two video programs to describe it. In a loose sense, Lesson 17 is based on what we have already learned about the early history of the universe, whereas Lesson 18 is devoted to those things we are still actively trying to find out. But the concepts in these two lessons are deeply intertwined.

For the longest time, under the scrutiny of astronomers, the universe seemed eternal and changeless. Newborn stars might condense out of interstellar material even as other older stars were fading (or blowing!) away. But on a larger scale, not much seemed to vary, particularly when compared to the minute lifetimes of individual astronomers or of modern astronomy itself. Even when galaxies were discovered at the start of the Roaring Twenties, it was presumed they came and went endlessly, as the stars within them did. There was no sense that the universe had either a beginning or an end.

But that idea crashed, along with the stock market, in 1929. In that year, Edwin Hubble discovered the relationship between galactic red shifts and distances, which was described briefly in Lesson 15. From that seemingly obscure observation, we suddenly had a reason to think that the universe was expanding, that it had a beginning only several times older than the age of Earth, and above all that it was evolving. That means the early universe was a very different place than it is today.

Learning Objectives

When you have completed all assignments for this lesson, you should be able to:

1. Define Olbers' paradox and describe its resolution. *HORIZONS* TEXTBOOK PAGES 296–298; VIDEO PROGRAM.

2. Identify and define the basic assumptions of cosmology, such as isotropy, homogeneity, and the resulting cosmological principle. *HORIZONS* TEXTBOOK PAGES 305–306.

3. Explain how the Hubble law and general theory of relativity predict the expansion of the universe and the geometry of space-time. *HORIZONS* TEXTBOOK PAGES 298–301 AND 306–309; VIDEO PROGRAM.

4. Explain the discovery and latest observations of primordial background radiation and show how these findings support the big bang theory and contradict the steady state theory. *HORIZONS* TEXTBOOK PAGES 301–302; VIDEO PROGRAM.

5. Recount currently accepted theories about the early history of the universe from the big bang until the beginning of the formation of galaxies. *HORIZONS* TEXTBOOK PAGES 302–305; VIDEO PROGRAM.

6. Describe clusters, superclusters, filaments, voids, and other large-scale structures in the universe and their implications for the process of galactic formation. *HORIZONS* TEXTBOOK PAGES 317–319; VIDEO PROGRAM.

Viewing Notes

"The Big Bang" video program begins by considering Olbers' paradox, which deals with the easily asked, though profoundly answered question: "Why is the sky dark at night?" If the universe were infinite in size and age, and filled uniformly with stars of essentially constant distances from us, then the sky would be very bright. So, at least one of these assumptions is wrong, and the principal error turned out to be the notion that the universe is eternal. As long as stars have not always existed, then the light from the most distant stars would not have had time to reach us yet.

The first inkling of this occurred when astronomers coupled Hubble's red shift-distance relation with Einstein's General Theory of Relativity and came to the conclusion that we live in a universe in which space-time itself is expanding. Relativity allows the universe to be either finite or infinite in extent (that's one of the "details" we are still trying to find out), but either way it is expanding and has neither a center nor an edge.

These are reasonably abstract concepts, but as the next segment explains, they led to one exceedingly important conclusion: that our expanding universe was extremely hotter and denser in the finite past, and that our entire universe, time and space, matter and energy, and all the laws of nature, may have come into existence at a single pivotal moment known as "time equals zero" or, more fondly, the big bang. For a while, some astronomers offered a competing theory that dismissed the restrictions of General Relativity and held that the universe was in an eternal, steady state. But later observations—including the momentous discovery by Arno Penzias and Robert Wilson, two physicists at the Bell Laboratories, who discovered cosmic microwave background radiation in 1964—laid the steady state theory to rest.

That radiation is described in the next segment. An infrared/microwave satellite called COBE, for COsmic Background Explorer, made more precise measurements of the microwave background. Not only is it thermal—perfect black-body radiation that shows the universe was indeed hotter and denser in the past—but COBE also found huge areas of the universe to be ever so slightly hotter and denser, by mere millionths of a degree, than others. That means the universe was already beginning to organize itself by the time its expanding, thinning material became transparent.

Today, there is overwhelming evidence that a big bang occurred, so the lesson concludes by describing the many critical stages that the universe passed through during the first million or so years of its approximately 14-billion-year history. But that means there are still many questions left unanswered. Exactly how long ago was $t = 0$? How did the universe change so quickly from the uniform gas of the microwave background epoch to the galaxies, clusters of galaxies, superclusters, filaments, and voids that we see today?

And why does the universe seem to be balanced right on the brink between perpetual expansion and nature-crunching collapse, like flipping a coin to see if it lands heads or tails but having it land on its edge instead? Lurking behind all of these questions is the ultimate one: What is the nature of the invisible dark matter that constitutes the vast bulk of the universe and stealthily guides its behavior? For all these reasons, there is Lesson 18.

As you watch the video program, consider the following questions:

1. Why *is* the sky dark at night, assuming you do not live in a big city?

2. What reasons do we have to think that the universe is expanding?

3. What reasons do we have to think that there was a big bang?

4. What is believed to have taken place in the few hundred thousand years following the big bang?

5. What is the origin of the cosmic microwave background radiation? Why does it correspond to a temperature of less than 3 degrees Kelvin?

6. How has information obtained from the COBE satellite contributed to cosmological research?

(The final segment of video program 15, "Galaxies," which may usefully be reviewed for this lesson, describes the role of gravity in binding clusters into the large-scale structure of the universe and how galaxies stream through space in response to it.)

Review Activities

Matching

Match the terms listed below with the definitions that follow. Check your answers with the Answer Key and review any terms you missed.

_____ 1. cosmology

_____ 2. Olbers' paradox

_____ 3. homogeneity

_____ 4. isotropy

_____ 5. universality

_____ 6. cosmological principle

_____ 7. closed universe

_____ 8. flat universe

_____ 9. open universe

_____ 10. big bang theory

_____ 11. cosmic microwave background radiation

_____ 12. steady state theory

_____ 13. recombination

a. theory that the universe does not evolve

b. theory that the universe is expanding from a very hot, dense state

c. study of the nature, origin, and structure of the universe

d. assumption that any observer in any galaxy sees the same general features in the universe

e. present appearance, at 2.7 degrees Kelvin, of the hot clouds of the big bang

f. stage, within 1 million years or less of the big bang, at which the material in the universe became transparent to radiation

g. apparent conflict about why the sky is dark at night

h. situation in which the average density of the universe is less than that necessary to halt its expansion

i. situation in which the space-time of the universe is not curved

j. situation in which the expansion of the universe will halt

k. assumption that the general features of the universe look the same in every direction

l. assumption that, on the largest scale, matter is uniformly spread throughout the universe

m. assumption that physical laws on Earth apply everywhere

Completion

Fill each blank with the most appropriate term from the list for that paragraph. A term may be used once, more than once, or not at all. Check your answers with the Answer Key and review when necessary. If a question requires two or more answers in succession, they may be in any order, unless the question indicates otherwise.

1. The cosmological principle is the assumption that the features seen by any observer, anywhere in the universe, appear _____. This principle is based on two basic assumptions. One assumption, that the average density of matter in space, when viewed on the _____ scale, is the same everywhere, is _____. The other assumption, that the universe on the _____ scale, looks _____ in every direction is known as _____. The assumptions of the cosmological principle are consistent with the theory that the universe began in a high-temperature, high-density state, known as the _____ theory.

big bang	largest
different	smallest
homogeneity	steady state
isotropy	the same

2. A closed universe is described as _____ and _____. The curvature of a closed universe would be _____, and the area of a very large circle would be too _____. An open universe is described as _____ and _____. The curvature of an open universe would be _____, and the area of a very large circle would be too _____. A flat universe is described as _____ and _____. The curvature of a flat universe would be _____, and the area of a very large circle would be _____. Regardless of the type of universe, no universe has an _____ or a _____.

bounded	negative
center	normal
closed	open
edge	positive
finite	small
flat	unbounded
infinite	zero
large	

Self-Test

Select the one best answer for each question. Check your answers with the Answer Key and review when necessary.

1. The main reason the night sky is dark is that the universe is
   a. static.
   b. infinite in size.
   c. finite in age.
   d. expanding.

2. Based on Olbers' original assumptions, the night sky should have been
   a. dark.
   b. infinitely bright.
   c. as bright as the surface of an average star.
   d. about 90 percent the brightness of an average star.

3. Isotropy is the assumption that
   a. matter is uniformly spread throughout space.
   b. the universe looks the same in every direction.
   c. the physical laws we know on Earth apply everywhere in the universe.
   d. the universe is expanding.

4. The cosmological principle allows for the universe to
    a. have a center.
    b. have an edge.
    c. look very different at different times.
    d. look very different in different directions.

5. According to the Hubble law, if Galaxy A is 10 times more distant than Galaxy B, then Galaxy A is
    a. receding 10 times faster than Galaxy B.
    b. approaching 10 times faster than Galaxy B.
    c. receding 100 times faster than Galaxy B.
    d. approaching 100 times faster than Galaxy B.

6. In which kind of universe would the area of even a very large circle be exactly equal to πr^2 (pi times the square of the circle's radius)?
    a. in every kind of universe
    b. closed universe
    c. open universe
    d. flat universe

7. The primordial background radiation is observable primarily at
    a. radio wavelengths.
    b. ultraviolet wavelengths.
    c. X-ray wavelengths.
    d. gamma ray wavelengths.

8. Detection of primordial background radiation
    a. supports the big bang theory.
    b. supports the steady state theory.
    c. contradicts the big bang and steady state theories.
    d. contradicts the big bang theory.

9. The photons we see as the primordial background radiation were emitted approximately

   a. 0.0001 seconds after the big bang.

   b. 4 seconds after the big bang.

   c. 3 minutes after the big bang.

   d. 300,000 years after the big bang.

10. Thirty minutes after the big bang, the second most abundant chemical element in the universe, after hydrogen, was

    a. lithium.

    b. helium.

    c. carbon.

    d. beryllium.

11. Of the following groupings of galaxies, the one that is largest in size is the

    a. Local Supercluster.

    b. Local Group.

    c. walls and filaments.

    d. Coma Cluster.

12. Galaxies and clusters of galaxies are distributed in space in superclusters, filaments, walls, and voids that make up the enormous groupings that astronomers call "structure." The origin of this structure is thought to be _____, which provides the gravitation and _____, which provide nuclei around which structure grows.

    a. neutron stars; supernovae

    b. dark matter; quantum fluctuations

    c. black holes; white dwarfs

    d. brown dwarfs; protostars

Lesson Review
Lesson 17
The Big Bang

Please Note: Use this matrix to guide your study and achieve the learning objectives of this lesson. It will also help you to view the video, which defines and demonstrates important concepts and skills as they relate to everyday life.

Learning Objective	Textbook	Telecourse Student Guide	Background Notes
1. Define Olbers' paradox and describe its resolution.	pp. 296–298	Matching Activities: 1, 2; Self-Test Questions: 1, 2.	
2. Identify and define the basic assumptions of cosmology such as isotropy, homogeneity, and the resulting cosmological principle.	pp. 305–306	Matching Activities: 3, 4, 6; Completion Activities: 1; Self-Test Questions: 3, 4.	
3. Explain how the Hubble law and general theory of relativity predict the expansion of the universe and the geometry of space-time.	pp. 298–301, 306–309	Matching Activities: 5, 7, 8, 9, 10; Completion Activities: 2; Self-Test Questions: 5, 6.	
4. Explain the discovery and latest observations of primordial background radiation and show how these findings support the big bang theory and contradict the steady state theory.	pp. 301–302	Matching Activities: 11, 12; Self-Test Questions: 7, 8.	
5. Recount currently accepted theories about the early history of the universe from the big bang until the beginning of the formation of galaxies.	pp. 302–305	Matching Activities: 13; Self-Test Questions: 9, 10.	
6. Describe clusters, superclusters, filaments, voids, and other large-scale structures in the universe, and their implications for the process of galactic formation.	pp. 317–319	Self-Test Questions: 11, 12.	

18

The Fate of the Universe

Assignments

For the most effective study of this lesson, we suggest that you complete the assignments in the sequence listed below.

Before Viewing the Video Program

- Read the Overview and Learning Objectives for this lesson. Use the Learning Objectives to guide your reading, viewing, and thinking.

- Review Chapter 15, pages 295–321, in the *Horizons* textbook, especially "Model Universes," pages 306–307, "Dark Matter in Cosmology," pages 307–312, and section 15-3, "21st-Century Cosmology," pages 312–319.

- Read Background Notes 18, "A Twenty-First Century Postscript," following Lesson 18 in this guide.

- Read the Viewing Notes in this lesson.

View "The Fate of the Universe" Video Program

After Viewing the Video Program

- Briefly note your answers to the questions listed at the end of the Viewing Notes.

- Review all reading assignments for this lesson, especially the Chapter 15 summary on pages 319–320 in *Horizons* and the Viewing Notes in this lesson.

- Respond to the challenges presented by each Critical Inquiry in the reading assignment and write brief answers to review questions 9–12 at the end of Chapter 15 in *Horizons* to be certain you understand the text material.

- Complete the Review Activities in this guide to reinforce your understanding of important terms and concepts. Check your answers with the Answer Key and review when necessary.

- Take the Self-Test in this guide to measure your achievement of the Objectives. Check your answers with the Answer Key and review when necessary.

- Complete any other activities and projects assigned by your instructor.

Overview

The previous lesson described our universe in its infancy, a time when there was little more than hydrogen and helium vapor slowly drawing itself together with its gravity to form the young galaxies. The universe was not yet filled with the stars and, presumably, planets and life-forms that tempt us to explore it today. But as much as we've learned about those earliest moments, there are some very fundamental questions left to be answered—questions that though still *un*answered, are nevertheless answerable. We have only to make the necessary observations, yet have been thwarted mostly by the fact that we must study objects at very great distances in order to see them as they were in the distant past.

For example, we have yet to discern why the space-time of the universe is so flat, and why the initial distribution of material within it so smooth. The answer probably lies in events that took place less than a second after the beginning, when the whole universe played by the rules of quantum theory that, today, govern only the tiny spaces inside of atoms.

We don't yet know how this smooth gas organized itself into the frothy large-scale structure that exists today. It seems that it happened all too quickly, but that perception is complicated by the fact that we don't know the age of the universe with certainty, nor do we know the nature of the dark matter that makes up 90 to 99 percent of the universe and would have contributed to the organizational process in unknown ways.

And until we understand that dark matter, we probably will not be able to determine the fate of the universe, for density equals destiny. The amount of material and consequent gravity determines whether the expansion will halt or persist forever. In the former case, the universe may collapse, perhaps to trigger another big bang in an endless series of recycling universes. In the latter, the universe happened only once. It may exist forever more, but the consumption of usable energy would eventually leave it a stagnant place. We are fortunate to inhabit the universe while it is young.

Learning Objectives

When you have completed all assignments for this lesson, you should be able to:

1. Identify unsolved problems in cosmology and describe recent attempts to refine the big bang theory with quantum concepts such as an inflationary universe and Grand Unified Theories (GUTs). *HORIZONS* TEXTBOOK PAGES 312–314; VIDEO PROGRAM.

2. Briefly relate several possible courses for the future of the universe and the kinds of observations that would be necessary to resolve the issue. *HORIZONS* TEXTBOOK PAGES 306–307; VIDEO PROGRAM.

3. Describe recent attempts to determine the age of the universe and the contradictions encountered between different techniques. *HORIZONS* TEXTBOOK PAGES 314–317; VIDEO PROGRAM.

4. Describe recent attempts to establish the existence and nature of dark matter and explain the implications of dark matter for the origin of the universe's large-scale structure and galaxies. *HORIZONS* TEXTBOOK PAGES 307–312 AND 317–319; VIDEO PROGRAM.

Viewing Notes

Using the poetry of Robert Frost, this video program opens with cosmology's most perplexing question: Will the universe end in the icy coldness of perpetual expansion, or in the fiery heat of some future collapse? The answer depends both on its rate of expansion and on the average density of its matter, because that determines the amount of gravity that the universe has available to halt itself.

But as the first segment explains, most of the material in the universe is unrecognizable to us. This dark matter, once erroneously identified as "missing mass," can be studied with gravitational lenses and other techniques. But we don't know exactly how much there is or *what* it is. It comprises at least 90 percent of the matter that exists and, if that's all there is, it could be "baryonic." That is, it could be ordinary matter with which we are familiar, but locked in the form of distant planets and failed or dead stars that are too faint to detect. But it could also represent as much as 99 percent of the universe's mass. In that case, at least some of it must be an exotic form of matter, such as massive neutrinos or some even stranger particles. Although the video describes a laboratory experiment that will attempt to detect such exotic material, its success cannot be assured.

The next segment describes the other unknown—the expansion rate that is represented by a number called the Hubble constant. This rate determines not only the fate of the cosmos, but its age: the elapsed time since the big bang. The best measurements of the Hubble constant were made by the Hubble Space Telescope (HST). It was named after Edwin Hubble not only to honor him but also because it is literally continuing what was his life's work: the observation of Cepheid variable stars in the most distant galaxies possible. Early results place the universe significantly younger than the 15 billion years quoted throughout the video lessons (which were produced just before the release of the first HST data). But there is still much room for error. In particular, the video describes the cosmological constant, an idea first proposed by Einstein and then rejected by him as unnecessary. It represents a repulsive force in the space between galaxies, which counters their gravity and can accelerate their expansion. If it actually exists, the universe could be older than the Hubble constant suggests.

The final segment dwells on a particular irony involving the relation between the fate of the universe and the curvature of space and time within it. If a universe that easily expands forever has an open curvature (shaped rather like an infinite saddle) and one that quickly arrests its expansion and collapses has a closed curvature (like a finite sphere), then it is remarkable how *flat* our own universe actually is. Much more simply, our measurements have not been precise enough to determine if the universe is open or closed, because the universe is agonizingly close to the borderline between those two possibilities. The video suggests that there may be a reason for the universe's flatness and describes a possible period of extremely brief but rapid inflation immediately following the big bang.

It was inevitable that our search for the origin and the future of everything would leave us with such uncertainty and speculation, but the course will next shift gears and zoom in upon our solar system, which the space age has made an increasingly familiar backyard in space.

As you watch the video program, consider the following questions:

1. How is the Hubble Space Telescope used in cosmological research?

2. How do we go about determining the age of the universe?

3. What reasons do we have to conclude that the universe contains a substantial amount of dark matter?

4. What are some of the forms that dark matter may take?

5. How are gravitational lenses and other techniques used to study dark matter?

6. What is meant by an inflationary universe?

7. What factors determine the fate of the universe, and what observations must we make to learn it?

Review Activities

Matching

Match the terms listed below with the definitions that follow. Check your answers with the Answer Key and review any terms you missed.

____ 1. critical density   ____ 5. inflationary universe

____ 2. oscillating universe theory   ____ 6. flatness problem

____ 3. horizon problem   ____ 7. Grand Unified Theories (GUTs)

____ 4. dark energy   ____ 8. hot or cold dark matter

a. the peculiar circumstance that the early universe appears to have had exactly the right amount of matter to make space-time flat

b. the circumstance that the primordial background radiation seems more uniform than expected by the big bang theory

c. the quality the distribution of matter in the universe must have in order for space-time to be flat

d. a version of the big bang theory, derived from GUTs, in which the horizon and flatness problems are resolved by a predicted event within the first second after the big bang

e. a situation that is possible only if the universe has greater than critical density

f. nonluminous matter detectable only by its gravity and moving either very fast or very slowly, respectively

g. attempts to describe the electromagnetic, weak, and strong forces of nature in a similar way

h. energy believed to fill empty space and accelerate the expansion of the universe

Completion

Fill each blank with the most appropriate term from the list for that paragraph. A term may be used once, more than once, or not at all. Check your answers with the Answer Key and review when necessary.

1. If the density of the universe is exactly equal to the critical density, then the universe is said to be _____. If the universe has less than the critical density and expands forever, the universe is said to be _____, but if the expansion halts at a finite future time, the universe is said to be _____. If the universe collapsed and experienced another big bang, it would be described as _____. If the universe experienced an extremely brief period of extremely rapid expansion immediately after the big bang, then it would be described as _____. Such a brief, rapid expansion is theorized to result from the _____ separating from the _____ and the _____.

closed	open
electromagnetic force	oscillating
flat	strong force
gravitational force	weak force
inflationary	

Self-Test

Select the one best answer for each question. Check your answers with the Answer Key and review when necessary.

1. Of the following combinations of types of universes, the one that is **NOT** possible is the
   a. closed, inflationary, and oscillating universe.
   b. open, inflationary, and oscillating universe.
   c. flat and inflationary universe.
   d. closed and oscillating universe.

2. The force that is **NOT** unified with the others in a GUT is the
   a. strong force.
   b. electromagnetic force.
   c. weak force.
   d. gravitational force.

3. The inflationary universe theory predicts that the universe must be
   a. completely flat.
   b. nearly flat, but could be barely open or closed.
   c. very open.
   d. very closed.

4. The type of universe in which the average density is predicted to be greatest is the
   a. open universe.
   b. flat universe.
   c. closed universe.
   d. negatively curved universe.

5. If the Hubble constant equals 100 km/sec/Mpc, then the Hubble time equals
   a. 100 billion years.
   b. 20 billion years.
   c. 10 billion years.
   d. 1 billion years.

6. In the absence of a cosmological constant, the rate of expansion of the universe in the past should have been

   a. greater than the rate of expansion today.

   b. precisely the same as the rate of expansion today.

   c. less than the rate of expansion today.

   d. oscillating.

7. For any given value of the Hubble constant, the type of universe that would probably be the youngest is the

   a. open universe.

   b. flat universe.

   c. closed universe.

   d. steady state universe.

8. The most distant known galaxies had formed by the time the universe was no more than

   a. 0.01 percent of its present age.

   b. 1 percent of its present age.

   c. 7 percent of its present age.

   d. 25 percent of its present age.

9. The very tiny irregularities that COBE found all over the sky in the cosmic microwave primordial background radiation indicate that

   a. the Milky Way is moving relative to the background.

   b. the radiation is a perfect black-body.

   c. the gas produced by the big bang was perfectly uniform.

   d. large, slightly denser patches of gas had formed within the first few hundred thousand years after the big bang.

10. The percentage of the mass of the universe that is dark, whether as matter or energy, is estimated to be

    a. 1 to 2 percent.

    b. 10 to 20 percent.

    c. 50 percent.

    d. 90 to 99 percent.

Lesson Review

Lesson 18

The Fate of the Universe

Please Note: Use this matrix to guide your study and achieve the learning objectives of this lesson. It will also help you to view the video, which defines and demonstrates important concepts and skills as they relate to everyday life.

Learning Objective	Textbook	Telecourse Student Guide	Background Notes
1. Identify unsolved problems in cosmology and describe recent attempts to refine the big bang theory with quantum concepts such as an inflationary universe and Grand Unified Theories (GUTs).	pp. 312–314	Matching Activities: 3, 5, 6, 7; Completion Activities: 1; Self-Test Questions: 1, 2, 3.	
2. Briefly relate several possible courses for the future of the universe and the kinds of observations that would be necessary to resolve the issue.	pp. 306–307	Matching Activities: 1, 2; Completion Activities: 1; Self-Test Questions: 1, 2, 4.	
3. Describe recent attempts to determine the age of the universe and the contradictions encountered between different techniques.	pp. 314–317	Matching Activities: 4; Self-Test Questions: 5, 6, 7.	
4. Describe recent attempts to establish the existence and nature of dark matter and explain the implications of dark matter for the origin of the universe's large-scale structure and galaxies.	pp. 307–312, 317–319	Matching Activities: 8; Self-Test Questions: 7, 8, 9, 10.	

Unit III

Background Notes

A Twenty-First Century Postscript

The interviews for the video lessons were taped in the early 1990s and reflect then-current research and theories. Since they are updated regularly, the *Horizons* textbook and this student guide reflect new research and discoveries. Lesson 18 is the first lesson where information in the video is noticeably out of date. A decade ago, the biggest problem in cosmology was globular clusters that seemed to be older than the age of the universe itself. Astronomers reacted with a flurry of research to solve this contradiction.

More detailed models of stellar evolution eventually lowered the age of globular clusters to about 13 billion years. Yet, the Hubble Constant later revealed by the Hubble Space Telescope still left the universe younger than that as long as it was flat as it appeared to be. In the videos, astronomers anticipated this and spoke at length about a "cosmological constant." The existence of this repulsive force would reduce the gravitational deceleration of the universe's expansion and result in a somewhat older age.

While such speculation turned out to be prescient, it underestimated the effect. To their surprise, astronomers learned at the beginning of this decade that after a few billion years of initial deceleration, the expansion is now accelerating because the repulsion of the ever emptier space exceeds the attraction of the ever diminishing matter density. It was also a surprise that most of the dark matter is really dark energy.

Now the numbers add up. Just under 14 billion years old, the universe became transparent after a half million years, and the first stars began to shine after a couple hundred million years. Some of the still existing globular clusters formed before the universe was one billion years old. It appears we've answered Mr. Frost's question: the world will end in ice.

The interviews show how much scientific progress can be accomplished in ten years. Yet we still don't know the nature of non-baryonic cold dark matter or the detailed behavior of this repulsive force. Remember that abstract science forms the foundation of future technology. Although a locally weak repulsive force may never have any practical uses, the word "antigravity" raises interesting possibilities. Even as modern high technology grew from the unanticipated development of quantum physics a century ago, we may now be witnessing the first crude discoveries that will underlie future technologies.

LESSON
19

The Origin of the Solar System

Assignments

For the most effective study of this lesson, we suggest that you complete the assignments in the sequence listed below.

Before Viewing the Video Program

- Read the Overview and Learning Objectives for this lesson. Use the Learning Objectives to guide your reading, viewing, and thinking.

- Read Chapter 16, "The Origin of the Solar System," pages 322–345, in the textbook.

- Read the Viewing Notes in this lesson.

View "The Origin of the Solar System" Video Program

After Viewing the Video Program

- Briefly note your answers to the questions listed at the end of the Viewing Notes.

- Review all reading assignments for this lesson, especially the Chapter 16 summary on page 343 in *Horizons* and the Viewing Notes in this lesson.

- Respond to the challenges presented by each Critical Inquiry in the reading assignment and write brief answers to the review questions at the end of Chapter 16 in *Horizons* to be certain you understand the text material.

- Complete the Review Activities in this guide to reinforce your understanding of important terms and concepts. Check your answers with the Answer Key and review when necessary.

- Take the Self-Test in this guide to measure your achievement of the Objectives. Check your answers with the Answer Key and review when necessary.

- Complete the Using What You've Learned activities and any other activities and projects assigned by your instructor.

Overview

During the first part of this course, you learned about how astronomers practice astronomy. You learned about the instruments they use and how they analyze light to gain knowledge about the objects producing that light, including our own star, the sun. This knowledge provided a foundation for studying objects that exist far beyond our solar system. You learned about the different stars and the origin, evolution, and death of each type; about black holes; about galaxies and how they cluster; and about quasars. In the last two lessons, you became familiar with theories and observations about the origin and evolution of the universe itself. The course now returns closer to home and investigates celestial objects that, for many, are more fascinating and exotic than anything else in the universe. These objects are the members of our solar system: the planets, including Earth, and their moons, asteroids, and comets.

Some astronomers and many amateur astronomers find these relatively nearby objects so interesting for several reasons. First, these objects are the only celestial objects whose surfaces we can study in detail. Also, because some of the planets have geologic features not found anywhere on Earth, study of those features can help us to develop theories and models to explain the origin and evolution of these worlds. Second, on some of these other worlds, we can study the remnants of early planetary history, where no such evidence exists on Earth. Third, the planets enable us to study the effects of extreme temperatures, dense and rarefied atmospheres, and large and small gravitational fields. Study of these other worlds enables us to conduct experiments in an immense "laboratory" that cannot be duplicated on the surface of Earth. Finally, study of our solar system allows us to ask and possibly answer some of the most important questions posed by humans: How did Earth form? What influence, if any, did the members of our solar system have on the evolution of Earth and life itself? What can we learn about the future of Earth from studying the environments of the nearby planets and the motions of the smaller members of the solar system? Exploration of our solar system, analysis of the observations, and conclusions drawn from those observations provide us with fascinating views of the planets, their moons, and the other members of the solar system.

This lesson begins the exploration by explaining current knowledge and theory about the origin of the solar system. It covers the characteristics of the solar system and how theoretical models account for the observed characteristics. The video program for this lesson includes planetary scientists describing observations that have lead them to conclude that solar systems beyond our own exist.

Learning Objectives

When you have completed all assignments for this lesson, you should be able to:

1. State several reasons for studying the solar system and cite examples of how knowledge gained from such study has enhanced understanding of the formation of Earth. *HORIZONS* TEXTBOOK PAGES 323, 330–336, AND 336–341; VIDEO PROGRAM.

2. List the characteristic properties of the solar system and state current estimates of its age. *HORIZONS* TEXTBOOK PAGES 329–336 AND 340–343; VIDEO PROGRAM.

3. Outline the solar nebula theory and explain how the characteristic properties of the solar system provide evidence that supports that theory. *HORIZONS* TEXTBOOK PAGES 324–325 AND 326–343; VIDEO PROGRAM.

4. Describe the characteristics of the terrestrial and Jovian planets and describe and compare the processes that resulted in the formation of each. *HORIZONS* TEXTBOOK PAGES 329–333 AND 336–343; VIDEO PROGRAM.

5. Identify the minor members of the solar system and describe their characteristics. *HORIZONS* TEXTBOOK PAGES 329–334.

6. Outline theories on the formation of planetesimals and the growth of protoplanets. *HORIZONS* TEXTBOOK PAGES 338–341; VIDEO PROGRAM.

7. Describe the observations made and conclusions drawn in the search for extrasolar planets. *HORIZONS* TEXTBOOK PAGES 325–329.

Viewing Notes

"The Origin of the Solar System" opens with computer simulation providing a spectacular flyby of all the planets of our solar system. The video program then moves on to a detailed examination of the solar nebula theory that explains how our solar system was formed. While several astronomers comment on different parts of the theory, computer graphics illustrate the spinning solar nebula, the accretion of dust particles, and the formation of the protoplanetary disk.

The second segment, "Observation," examines how astronomers are searching distant stars to find evidence to support the solar nebula theory. At Kitt Peak, astronomer Frederick M. Walter describes the difficult search for signs of the formation of planets around distant stars in the Great Nebula in Orion, a region of active star formation. The video program also describes the work of radio astronomers in examining disks around new stars. Anneila Sargent, a radio astronomer at Owens Valley Radio Observatory, describes the conclusions she has drawn from her observations of protoplanetary disks, including the details she has seen around HL Tauri.

In searching for evidence of planetary formation, infrared telescopes have been especially valuable in detecting cooler dust disks around older stars. Richard Terrile, an astronomer at the Jet Propulsion Laboratory in Pasadena, California, describes the infrared image he captured of Beta Pictoris. That image shows a disk with a hollowed-out center approximately the size of our solar system; Terrile and other astronomers believe planets may be orbiting in the void around the star.

The last part of the video program focuses on the evolution of our solar system, beginning with descriptions of the rocky terrestrial planets and the gas giants. Planetary experts explain the role of heat and gravity in planetary formation and structure. They also describe cratering and explain why cratering is not evident on Earth's surface. Other astronomers explain the role of a planet's distance from the sun in determining its size and atmosphere and describe the concept of the "ice line" in the solar system. The video program also covers the various small bodies of the solar system, including asteroids and moons and concludes with a summary of the solar nebula theory and the implications of that theory for finding similar planetary systems around the other stars in our galaxy.

As you watch the video program, consider the following questions:

1. What is a protoplanetary disk and how is it formed?

2. What processes are involved in planetary formation?

3. What ends the process of planet formation and clears the disk?

4. How have dust disks around other stars been detected and what is the significance of the discovery of these dust disks?

5. What are the principal differences between the terrestrial and Jovian planets of our solar system?

6. What is the process that causes a terrestrial planet to form an iron core and a solid silicate crust?

7. What role did distance and temperature play in making some planets small, dense, and rocky and others large, gaseous, and low density?

8. What is the "ice line" and what role does it play in explaining the formation of the planets?

9. How does understanding the formation of our own solar system help us to understand the formation of solar systems around other stars?

Review Activities

Matching

Match the terms listed below with the definitions that follow. Check your answers with the Answer Key and review any terms you missed.

_____ 1. solar nebula theory   _____ 12. meteorite

_____ 2. gravitational collapse   _____ 13. condensation sequence

_____ 3. terrestrial planet   _____ 14. planetesimal

_____ 4. Jovian planet   _____ 15. differentiation

_____ 5. extrasolar planet   _____ 16. accretion

_____ 6. Galilean satellites   _____ 17. protoplanet

_____ 7. asteroids   _____ 18. half-life

_____ 8. comet   _____ 19. heat of formation

_____ 9. uncompressed density   _____ 20. radiation pressure

_____ 10. meteor   _____ 21. outgassing

_____ 11. meteoroid   _____ 22. heavy bombardment

a. planet with a large diameter and low density

b. an object that survives its passage through the atmosphere and strikes the ground

c. a rock in space before it enters Earth's atmosphere

d. energy released by infalling of matter during formation of a planetary body

e. theory that the planets and the sun formed from the same cloud of gas and dust

f. a small, dense, rocky planet

g. the sticking together of solid particles to produce a larger particle

h. order in which different materials condense from a gas in relationship to distance from the sun

i. density the planets would have if gravity did not compress them

j. small, icy body that orbits the sun

k. the time it takes for half the atoms of a radioactive element in a sample of material to decay

l. small bodies that formed from the solar nebula and eventually grew into protoplanets

m. the process that can only occur when a planet has grown to about 15 Earth masses, thus giving it enough gravity to begin capturing gas directly from the solar nebula

n. force exerted on the surface of a body by its absorption of light

o. separation of planetary material according to density

p. the four largest satellites of Jupiter

q. creation of a planet's atmosphere from the planet's interior

r. streak of light in the sky

s. minor planets, most of which orbit the sun between Mars and Jupiter

t. the process that cratered all of the solid worlds as the last debris in the solar nebula was swept up

u. massive object resulting from the coalescence of planetesimals in the solar nebula

v. planet orbiting a distant star (not the sun)

Completion

Fill each blank with the most appropriate term from the list for that paragraph. A term may be used once, more than once, or not at all. Check your answers with the Answer Key and review when necessary. If a question requires two or more answers in succession, they may be in any order, unless the question indicates otherwise.

1. In order to develop a theory that describes the origin of the solar system, the theory must be able to explain the major physical characteristics of the solar system. We can divide these characteristics into two main categories. The first category involves the rotational and orbital characteristics of the sun and planets; the second involves the physical properties of the planets and other minor members of the solar system.

    The sun rotates on its axis in approximately 30 days. If you were looking down on the solar system from the north (above), you would see that the direction in which the sun rotates is _____. If you were looking at the motion of the planets, you would observe that the rotation of most of the planets is _____. The only exceptions to this observation are the planets _____ and _____. Furthermore, all the solar system's _____ and almost all of the _____ orbit the sun in a _____ direction. These observations indicate that the cloud of material from which the solar system originated must have been rotating in the _____ direction.

The orbits of the planets tend to lie in a _____. Thus, if you view the solar system _____, the planets would appear to be in a _____ and _____ line on either side of the sun. If viewing the solar system from such a perspective, the shape would resemble a _____. Not only do the _____ of the planets lie within this relatively _____, but the _____ are only _____ tilted with respect to the plane of _____ orbit.

Based on their physical characteristics, the planets can be divided into _____ categories. The planets nearest the sun are very similar to planet _____. In size, these planets are relatively _____, their composition is _____, and their surfaces are _____. These planets are also referred to as the _____ planets. The size of the planets farthest from the sun is relatively _____, and their composition is _____. Also, we are unable to see _____ surfaces. These planets appear to be very similar to the planet _____ and therefore are referred to as the _____ planets. Their atmospheres contain the lightest gases, and _____.

Although the sun and planets are the major components of the solar system, any theory about the origin of the solar system must also account for the system's many minor members. These minor members include _____, in orbit around the major planets, _____, and _____. These minor objects are composed of various materials, including _____, _____, _____ and _____. In fact, one astronomer feels that the term "icy mud ball" should be used for _____. An additional and interesting observation of our solar system reveals that it is relatively empty of _____ and _____, which are remnants of the original cloud of material that formed the solar system. Taking all of these observations into account, astronomers, at present, think that the best explanation of the formation of our solar system is the _____.

asteroids	helium	rocky
clockwise	horizontal	same
comets	hydrogen	slightly
common plane	ice	small
counterclockwise	Jovian	solar nebula theory
disk	Jupiter	solid
dust	large	solid and rocky
Earth	minor members	straight
Earth's	moons	terrestrial
edge on	opposite	thin disk
equators	orbits	two
four	oxygen	Uranus
gas	planets	Venus
gaseous	rocks	

2. Planets begin to form when _____ join to make larger objects called _____. These objects grow large when they acquire additional material that collides with them. These objects do not shatter when they collide with this material because they are traveling in the _____ direction. Therefore, the relative _____ of the colliding objects are _____ and the materials stick to the larger object's surface. Adherence to the surface is also facilitated because of _____ and _____ on the surfaces of the smaller objects. Also, the adherence of larger _____ to these objects may be aided by _____, which helps to attract and hold additional material.

The formation of the terrestrial, or _____ planets, can be described by the following: Once the planet formed, _____, generated by the decay of _____ _____ elements, caused the interior of the planet to _____. As a result, _____ materials, such as _____ and _____, sank into the core of the planet. The lower density _____ rose to the surface of the planet forming a _____. This process of the separation of material according to density is called _____. During this process, the _____ atmosphere of Earth, probably consisting of _____ and _____, was driven off by this heat and outbursts from the young sun. In a process called

_____, new gases that escape from the melted rocks create a second atmosphere.

This theory of terrestrial planet formation can be improved in a couple of ways. Consider the _____ of the early solar nebula. Closest to the sun, temperatures were _____ and only particles of high melting point could solidify first. This physical environment caused particles _____ and _____ to cool first, and the protoplanets began by forming _____ cores. As the solar nebula continued to cool, the _____ condensed. The protoplanets then added silicate _____ and _____ to the already-formed metallic cores. The heat generated by the _____ matter, also referred to as the _____, aided in the melting of the planet's interior and differentiation. The planet differentiated as it formed.

This modification in the theory of formation of the terrestrial planets may also give additional clues to the formation of Earth's atmosphere. If Earth formed in a _____ state, then the _____ and _____ of the solar nebula may have never accumulated enough to become Earth's "first" atmosphere. The early atmosphere would have been formed by the process of _____, although not entirely. The _____ of formation would have prevented much _____ from forming in the atmosphere. It is believed that Earth's water and much of its present atmosphere was accumulated when Earth _____ with _____ planetesimals. These _____ planetesimals were probably formed in the _____ reaches of the solar system when encounters with the planet _____ drove them into the inner solar system and on a collision course with Earth.

atmosphere	helium	mantles	protoplanets
collided	high	melt	radioactive
crust	high-desity	metal oxides	same
crusts	high temperature	metallic	short-lived
differentiation		metals	silicates
Earthlike	hydrogen	molten	similar
electrostatic forces	icy	nickel	speeds
	infalling	original	sticky coatings
gravity	iron	outer	temperature
heat	Jupiter	outgassing	volatile-rich
heat of formation	low	planetesimals	water
	low-density		

Self-Test

Select the one best answer for each question. Check your answers with the Answer Key and review when necessary.

1. Evidence of cratering on the moon and the other terrestrial planets suggests that Earth

   a. never experienced a similar kind of cratering.

   b. experienced a similar kind of cratering earlier in its history.

   c. will experience a similar kind of cratering in the future.

   d. will never experience a similar kind of cratering.

2. Because of Earth's low mass, our planet's atmosphere

   a. contains mostly gases that are similar to those on the Jovian planets.

   b. contains the same percentages of gases that exist throughout the solar system.

   c. contains very little hydrogen and helium.

   d. is composed primarily of carbon dioxide.

3. The orbits of the planets
   a. all lie in approximately the same plane.
   b. lie perpendicular to Earth's orbit.
   c. are perfectly circular.
   d. are inclined to the ecliptic.

4. Each of the planets in our solar system orbits the sun
   a. in circular orbits.
   b. in the direction opposite to which they rotate, clockwise as seen from the north (above).
   c. in the same direction.
   d. opposite in direction to the orbital motion of Earth.

5. The orbits of the planets lie mainly in the same plane because the planets
   a. formed from a large thin disk of material surrounding the new sun.
   b. were captured by an already-formed sun.
   c. were pulled out of the sun by the gravitational influences of a star passing nearby.
   d. formed at a variety of locations but were pulled into a single plane by the gravitational influence of the sun.

6. The solar system has relatively little gas and dust between the planets. The lack of this material is probably due to the
   a. "vacuuming up" of the material by the gravitational attraction of the planets.
   b. condensation properties of the Jovian planets.
   c. radiation pressure from the young sun.
   d. accretion processes of the terrestrial planets.

7. The solar nebula theory describes how the planets formed from
   a. the capture of planetesimals in orbit around a nearby star.
   b. the condensation of a cloud of pure hydrogen and helium.
   c. the accretion of high-density matter in the vicinity of the newly forming sun.
   d. a disk-shaped cloud of gas and dust around the protosun.

8. The terrestrial planets nearest the sun are
   a. large and composed of ice and hydrogen and helium gas.
   b. small, solid, and rocky.
   c. irregular in shape and composed mostly of ice and rock.
   d. spherical in shape with solid crusts and gaseous interiors.

9. The Jovian planets are made primarily of
   a. hydrogen and helium.
   b. methane and ammonia.
   c. liquid hydrogen.
   d. icy rock in solid form.

10. Most asteroids revolve around the sun between the orbits of
    a. Jupiter and Saturn.
    b. Earth and Mars.
    c. Saturn and Uranus.
    d. Mars and Jupiter.

11. Objects made mostly of ice that form a large gaseous "tail" that is directed away from the sun at all times are called
    a. asteroids.
    b. meteors.
    c. meteorites.
    d. comets.

12. Formation of planetesimals by condensation refers to the process of
    a. meteorites colliding and sticking together.
    b. asteroids colliding and breaking apart.
    c. matter building one atom at a time.
    d. asteroid collisions and adhesion.

13. Solid particles sticking together to form larger particles describes the process of

   a. accretion.

   b. adhesion

   c. condensation.

   d. adherence.

14. The separation of low-density and high-density materials in a planet during the early stages of its formation is referred to as

   a. accretion.

   b. differentiation.

   c. homogeneous condensation.

   d. heterogeneous condensation.

15. The interiors of Jupiter and Saturn consist mostly of

   a. icy rock in solid form.

   b. hydrogen and helium.

   c. liquid hydrogen.

   d. methane and ammonia.

16. Low density and cold dust disks have been found around many stars. These disks are believed to be regions where planets have formed or are in the process of forming. The disks are most readily detected by

   a. infrared telescopes.

   b. optical telescopes.

   c. orbiting ultraviolet telescopes.

   d. X-ray sensitive satellite detectors.

17. A planet that has formed around the star 51 Pegasi has been detected by

   a. direct observation in an optical telescope.

   b. infrared sensors on the Hubble Telescope.

   c. Doppler shifts in the star's spectrum as the planet orbits, causing the star to wobble back and forth.

   d. no such planet has been observed.

Using What You've Learned

1. To understand the relative sizes of the planets, construct a scale model of the solar system. Start by establishing a scale for the diameters of the planets. Use Earth's diameter of 12,656 kilometers as equal to 1 inch. Then, using the data on "Physical Properties" in Table A-13, "Properties of the Planets," at the top of page 460 of the *Horizons* textbook, calculate inch equivalents for all the other planets. To translate these figures into representations of the sizes of the planets, take a roll of adding machine paper tape and cut it into appropriate lengths to represent the diameters of the other planets. For example, Jupiter has a diameter of 142,400 kilometers, which is 11.18 times the diameter of Earth. Cut a piece of tape 11.2 inches long. Place this tape next to the tape representing Earth. Only by placing the strips of paper next to each other can you appreciate the relative sizes of the terrestrial and Jovian planets.

   Complete the scale representation of the solar system by cutting a strip of paper representing the sun. The sun has a diameter of about 1,400,000 kilometers, compared to a diameter of 12,656 kilometers for Earth. Divide the sun's diameter by Earth's diameter and you discover that the sun is approximately 111 times larger in diameter than Earth. Cut a piece of tape 111 inches long and place that next to the pieces representing the planets.

2. Repeat the above activity, this time comparing the distances of the planets from the sun. Use the data on "Orbital Properties" in Table A-13, "Properties of the Planets," in the middle of page 460 of the *Horizons* textbook, to determine the distance of each planet from the sun. Use the scale of 1 foot to 1 astronomical unit. According to this scale, Earth is 1 foot away from the sun. Although you can quickly see that Pluto would be 39 feet (39 astronomical units) away from the sun, you cannot appreciate how far that actually is until you actually unroll a 39-foot strip of adding machine tape. Cut a piece of tape for each of the distances given in the Orbital Properties table in either textbook and compare. (Be careful not to compare the lengths of tape in exercise 1 with the lengths of tape in this exercise. These exercises did not use the same scale.)

Lesson Review

Lesson 19

The Origin of the Solar System

Please Note: Use this matrix to guide your study and achieve the learning objectives of this lesson. It will also help you to view the video, which defines and demonstrates important concepts and skills as they relate to everyday life.

Learning Objective	Textbook	Telecourse Student Guide	Background Notes
1. State several reasons for studying the solar system and cite examples of how knowledge gained from such study has enhanced understanding of the formation of Earth.	pp. 323, 330–336, 336–341	Self-Test Questions: 1, 2.	
2. List the characteristic properties of the solar system and state current estimates of its age.	pp. 329–336, 340–343	Matching Activities: 18; Completion Activities: 1; Self-Test Questions: 3, 4, 5.	
3. Outline the solar nebula theory and explain how the characteristic properties of the solar system provide evidence that supports that theory.	pp. 324–329, 336–343	Matching Activities: 1, 2, 9, 13, 20; Completion Activities: 1, 2; Self-Test Questions: 6, 7.	
4. Describe the characteristics of the terrestrial and Jovian planets and describe and compare the processes that resulted in the formation of each.	pp. 329–333, 326–343	Matching Activities: 3, 4, 6, 22; Completion Activities: 1, 2; Self-Test Questions: 8, 9, 15.	
5. Identify the minor members of the solar system and describe their characteristics.	pp. 329–331, 334	Matching Activities: 7, 8, 10, 11, 12; Completion Activities: 1; Self-Test Questions: 10, 11.	
6. Outline theories on the formation of planetesimals and the growth of protoplanets.	pp. 338–341	Matching Activities: 14, 15, 16, 17, 19, 21; Completion Activities: 2; Self-Test Questions: 12, 13, 14.	

Learning Objective	Textbook	Telecourse Student Guide	Background Notes
7. Describe the observations made and conclusions drawn in the search for extrasolar planets.	pp. 325–329	Matching Activities: 5; Completion Activities: 1; Self-Test Questions: 16, 17.	

LESSON
20

Planet Earth

Assignments

For the most effective study of this lesson, we suggest that you complete the assignments in the sequence listed below.

Before Viewing the Video Program

- Read the Overview and Learning Objectives for this lesson. Use the Learning Objectives to guide your reading, viewing, and thinking.
- Read Chapter 17, "The Earthlike Planets," section 17-1, pages 346–352, in the *Horizons* textbook.
- Read the Viewing Notes in this lesson.

View the "Planet Earth" Video Program

After Viewing the Video Program

- Briefly note your answers to questions listed at the end of the Viewing Notes.
- Review all reading assignments for this lesson, especially the Chapter 17 summary page 378 in *Horizons* and the Viewing Notes in this lesson.
- Respond to the challenges presented by each Critical Inquiry in the reading assignment and write brief answers to review questions 1–4 at the end of Chapter 17 in *Horizons* to be certain you understand the text material.
- Complete the Review Activities in this guide to reinforce your understanding of important terms and concepts. Check your answers with the Answer Key and review when necessary.
- Take the Self-Test in this guide to measure your achievement of the Objectives. Check your answers with the Answer Key and review when necessary.

- Complete the Using What You've Learned Activities and any other activities and projects assigned by your instructor.

Overview

The previous lesson began the study of the solar system—the region of the universe that we would consider to be local. Our solar system consists of several types of objects: the sun, nine major planets, several dozen moons, and a multitude of smaller objects such as comets and asteroids. Planetary astronomers categorize the nine planets into two groups according to their similar surface features and internal characteristics. This lesson is the first of three that examine the terrestrial planets, beginning with Earth. (The Jovian planets are covered in Lessons 23 and 24.) The terrestrial planets have relatively high densities, hard surfaces, and few satellites. In order to understand the formation and evolution of these planets, we must compare their characteristics and history to Earth's. The more we know about Earth, the better we will be able to understand the forces that shaped the other terrestrial planets. This program is devoted to the planet we have studied for the longest time and in the greatest detail. The knowledge we gained and the processes we have employed in studying Earth will be used to study the other planets. Thus, this lesson provides the foundation for the study of the rest of the solar system and the application of what is called *comparative planetology*.

When Earth was formed by the accretion process, the heat generated by these impacts and the heat released as a result of the decay of radioactive elements caused the planet's interior to become molten. While Earth was in this molten state, heavier elements such as nickel and iron drained to the center, while the lighter silicates rose to the surface. These lighter compounds solidified and became Earth's crust. This process of differentiation, the separation of materials into layers depending on their density, is common to all the terrestrial planets. Earth, however, is unique among the planets in that it is the only planet that continues to have extensive activity in its solidified crust.

As this lesson explains, the heat in the middle layer, or mantle, of Earth keeps the material underneath the crust in constant motion, causing the crust to crack and form large sections. These sections, or plates, collide and slide past each other, producing earthquakes, volcanoes, mountain ranges, and deep ocean trenches. No other planet has a geology so driven. However, as we learn more about Earth's geologic activity, we also come closer to understanding why the other planets do not show the same plate tectonics.

Erosion and the motion of Earth's crust through plate tectonics have destroyed all signs of the planet's earliest geologic record. No crater remains that is more than a few tens of thousands of years old. Plate tectonics are also responsible for the development of Earth's atmosphere. Gases such as carbon dioxide and water vapor, released from the molten interior, altered the history of Earth. Water vapor, in an environment cool enough for it to condense and form liquid water, removed the carbon dioxide from the atmosphere and allowed Earth's surface to cool.

Chemicals within that liquid water began to form complex molecules and duplicate themselves. Life had formed. Once this life was capable of photosynthesis, it produced extra oxygen, eventually allowing a layer of ozone to form in the upper atmosphere. This ozone layer protected the primitive life-forms from harmful radiation and, in time, allowed life to move onto the land. Earth's atmosphere, from its earliest composition of gases left over from the solar nebula to its present composition of oxygen and nitrogen, is dynamic and critical in the support of life on this planet.

Today, however, the atmosphere that protected primitive life in the oceans and on land may be in danger because of the industrial processes in which that life is engaged. Carbon dioxide, once removed by liquid water, is now being returned to the atmosphere, possibly raising global temperatures. The ozone layer, screening harmful ultraviolet radiation, is being destroyed by chlorofluorocarbons (CFCs). The geological activity on Earth will continue regardless of the activity of the life that currently inhabits this planet. However, if intelligent life is to survive on this planet, we must understand the geological and atmospheric processes that affect it. As we study Earth to better understand the other terrestrial planets, we see that by studying the other terrestrial planets we can better understand processes that occur on Earth. In addition to gaining knowledge about the individual objects in our solar system, the combined knowledge helps us to understand the origin of the solar system itself.

Learning Objectives

When you have completed all assignments for this lesson, you should be able to:

1. Outline a four-stage history of Earth's formation. *HORIZONS* TEXTBOOK PAGES 346–347.

2. Describe a model of Earth's interior and explain how we have come to develop that model. *HORIZONS* TEXTBOOK PAGES 347–348; VIDEO PROGRAM.

3. Explain the current theory about how the magnetic field of Earth is produced. *HORIZONS* TEXTBOOK PAGE 347; VIDEO PROGRAM.

4. Explain the theory of and evidence for plate tectonics and continental drift. *HORIZONS* TEXTBOOK PAGES 348, 350–351; VIDEO PROGRAM.

5. Describe and explain the processes that have changed Earth's atmosphere from its earliest formation to the present day. *HORIZONS* TEXTBOOK PAGES 348–349, 352; VIDEO PROGRAM.

6. Explain the cause and possible consequences of a depletion of Earth's ozone layer. *HORIZONS* TEXTBOOK PAGES 349, 352; VIDEO PROGRAM.

7. Explain the concept of global warming and describe the possible consequences. *HORIZONS* TEXTBOOK PAGES 348–349; VIDEO PROGRAM.

8. Describe the effects life on Earth may have on Earth's environment. *HORIZONS* TEXTBOOK PAGES 349, 352 ; VIDEO PROGRAM.

Viewing Notes

In this video program, planetary scientists, geophysicists, and geologists outline the formation and evolution of Earth. They also discuss how our understanding of the process that formed our planet contributes to our understanding of the other planets of our solar system and how, as we learn more about those other planets, our knowledge of Earth grows.

The first segment describes the early evolution of Earth. Computer animation illustrates the formation of the planet by accretion and its eventual differentiation into a sphere with an iron core surrounded by a mantle and crust. Planetary scientists describe the dynamo effect, the process by which the motion of the molten iron in the core generates Earth's magnetic field and how earthquakes and data gathered from seismometers have helped planetary scientists learn more about Earth's interior. As explained in the video program, earthquakes occur when plates "floating" on the mantle move. The motion is a result of convection, which causes the mantle material to flow continuously.

The video program then considers plate tectonics, one of the dynamic phenomena still shaping Earth today. Computer graphics illustrate how convection broke up the supercontinent of Pangaea and ultimately formed the continents as we know them today. This segment also explains the dynamic phenomena of plate tectonics, seafloor spreading, midocean ridges, subduction zones, and volcanism and their ongoing role in shaping Earth. Geophysicists and a planetary scientist briefly compare Earth to the other terrestrial planets, explaining the reason for the relatively small number of visible craters on Earth and noting the apparent absence of plate tectonics on other planets.

The video program also covers Earth's most unique features—its liquid water and atmosphere—and describes how water was formed and how our atmosphere evolved. This last segment of the program explains the changes that occurred in Earth's oceans and atmosphere that enabled life to develop on the planet. An atmospheric scientist and a geological oceanographer describe the role of carbon dioxide in Earth's atmosphere and oceans and how the oceans, over millions of years, dissolved much of the carbon dioxide, forming carbonates and carbonate rock (limestone). This segment also describes the greenhouse effect and the ozone layer and their role in allowing life to develop on Earth, and it explains how the oceans and atmosphere evolved to contain sufficient oxygen to allow organisms to live on land.

As you watch the video program, consider the following questions:

1. How did Earth form and what processes shaped it to its present form?

2. How are earthquakes and seismometers used to study the structure of Earth's interior?

3. How do plate tectonics reflect convection in the mantle and how are plate tectonics related to continental drift and seafloor spreading?

4. Why is more cratering visible on the other terrestrial planes and on the moon than on Earth?

5. What characteristics make Earth's surface unique among the terrestrial planets? Why?

6. How did Earth's atmosphere develop?

7. What changes occurred in Earth's atmosphere and oceans that enabled organisms to live on land?

8. What is ozone and why is it necessary in order for life to exist on Earth's surface?

Review Activities

Matching

Match the terms listed below with the definitions that follow. Check your answers with the Answer Key and review any terms you missed.

_____	1. comparative planetology	_____	8. rift valley
_____	2. mantle	_____	9. plate tectonics
_____	3. plastic	_____	10. greenhouse effect
_____	4. dynamo effect	_____	11. primeval atmosphere
_____	5. basalt	_____	12 Pangaea
_____	6. midocean rise	_____	13. seafloor spreading
_____	7. midocean rift	_____	14. continental drift

a. name of the single land mass from which the current continents were formed

b. a material with the properties of a solid but capable of flowing under pressure

c. layer of dense rock and metal oxides between the molten core and Earth's surface

d. process by which an atmosphere rich in carbon dioxide entraps heat and raises the temperature of a planetary surface

e. constant destruction and renewal of Earth's surface resulting from movement of sections of crust

f. observed motion of the large land masses sitting on top of the crustal plates as a result of plate tectonics

g. Earth's first air

h. dark igneous rock characteristic of solidified lava

i. study of planets in relation to one another

j. chasm that splits the midocean rises where crustal plates move apart

k. long, straight, deep valley produced by the separation of crustal plates

l. process by which Earth's magnetic field is generated in the conducting material of its molten core

m. process by which molten material from the mantle emerges from the midocean rifts, pushing the continents further apart

n. undersea mountain range that pushes up from the seafloor in the center of an ocean

Completion

Fill each blank with the most appropriate term from the list for that paragraph. A term may be used once, more than once, or not at all. Check your answers with the Answer Key and review when necessary. If a question requires two or more answers in succession, they may be in any order, unless the question indicates otherwise.

1. The crust of Earth is divided into numerous plates. Movement of these plates, which results from _____ in Earth's _____, causes the _____ to separate. Evidence of this separation is indicated by the presence of mountain ranges, or _____, in the middle of the oceans. In the center of these mountain ranges are _____, from which_____ rock rises up, causing the seafloor to _____ outward. The strongest evidence supporting the phenomenon of _____ is found in the residual _____ in the solidified rock on the seafloor. Some of this material is found to have a _____ polarity _____

from Earth's present magnetic field. These regions, on both sides of the _____, show _____, parallel, and _____ bands of magnetism that demonstrate that Earth is creating new _____, causing the _____ to spread and push the continents apart.

alternating	midocean rises
continents	molten
convection	reversed
crust	seafloor
magnetic	seafloor spreading
magnetism	spread
mantle	symmetric
midocean rifts	

2. Earth's first atmosphere, sometimes called the _____ atmosphere, was thought to consist of gases originally belonging to the _____. These original gases included _____, _____, _____, and _____. More recent studies of the origin of Earth, however, indicate that temperatures on the newly forming Earth were hot enough to release gases such as _____ and _____, thus permitting the first atmosphere to be composed of these gases. During the _____ stages of planet building, planetesimals rich in _____ vaporized materials such as _____, _____, and _____ collided with Earth and may have contributed a significant portion of Earth's early atmosphere. The atmosphere now rich in _____, _____, and _____ cooled, and these gases _____. As liquid water collected on Earth's surface, the water began to dissolve the _____, thus reducing the _____ and lowering Earth's surface _____. Earth's atmosphere continued to change. The primeval gases of _____ and _____ were destroyed by _____ radiation. With the reduction of the gas _____ being dissolved into the oceans, the atmosphere of Earth became rich in the gas _____. Another important gas that allowed living organisms to venture out of the oceans and onto the land eventually appeared in Earth's atmosphere. This gas, known as _____, prevented ultraviolet radiation from reaching Earth's surface.

Today, we face two possible problems because of human activity on Earth. First, we are concerned that the amount of carbon dioxide being released into the atmosphere is increasing. Since the amount of _____ radiation leaving the surface of Earth would be reduced, the temperature of Earth's surface would _____.
If this pattern of _____ were to continue unabated, the surface temperature would be such that the ice caps would begin to _____ and ocean levels would begin to _____.
Second, we are concerned about the destruction of the _____ in Earth's atmosphere. Modern chemicals called _____ destroy the _____ when they leak into the atmosphere. If this layer were significantly reduced, the amount of _____ reaching the surface of Earth would _____ and have a detrimental effect on life on this planet.

ammonia	melt
carbon dioxide	methane
chlorofluorocarbons	nitrogen
condensed	ozone
decrease	ozone layer
easily	primeval
final	rise
freeze	solar nebula
global warming	temperature
greenhouse effect	ultraviolet
helium	ultraviolet radiation
hydrogen	water
increase	water vapor
infrared	

Self-Test

Select the one best answer for each question. Check your answers with the Answer Key and review when necessary.

1. Iron and nickel drained to the center of Earth, while at the same time, the lighter silicates rose to the surface, solidified, and formed a thin crust. This process is known as

   a. differentiation.

   b. sedimentation.

   c. liquefaction.

   d. distillation.

2. Which of the following sequences best describes the four-stage history of Earth's formation and evolution?

   a. differentiation, surface evolution, cratering, basin flooding

   b. condensation, cratering, basin flooding, differentiation

   c. differentiation, cratering, basin flooding, surface evolution

   d. cratering, condensation, accretion, surface evolution

3. Earth's core has a solid center and liquid outer parts. Evidence for the liquid component of the core comes from

   a. gravity studies by spacecraft.

   b. seismic wave studies.

   c. physical measurement by subterranean probes.

   d. oil well drillings.

4. Between the crust and liquid core lies a solid material that behaves as a plastic, flowing under high pressure. This material makes up Earth's

   a. mantle.

   b. crust.

   c. inner core.

   d. continental shelf.

5. Convection currents in Earth's core stir the molten liquid. The rotating Earth and rotation of Earth's core produce Earth's

   a. aurora.

   b. seismic activity.

   c. plate tectonics.

   d. magnetic field.

6. Earth's magnetic field is thought to result from the dynamo effect, which is

   a. motion in the liquid core that generates seismic waves.

   b. motion in the liquid core that generates electric currents.

   c. thermal cooling in the liquid core that generates temperature gradients.

   d. seismic convection in the liquid core that generates liquefaction.

7. Earth's crust is divided into large sections that slowly move about as the result of

   a. seismic activity in the crust.

   b. convection currents in the core.

   c. convection currents in the mantle.

   d. seafloor separation.

8. One of the best ways to locate the crustal plate boundaries is to identify

   a. large mountain ranges on the continents.

   b. lakes, major rivers, and oceans.

   c. the outline of the continental shelf.

   d. the locations of earthquakes.

9. Earth's primeval atmosphere probably consisted of all of the following gases **EXCEPT**

   a. oxygen.

   b. carbon dioxide.

   c. water vapor.

   d. nitrogen.

10. The carbon dioxide in Earth's early atmosphere was produced when large amounts of gases were released during the accretion process when Earth's surface temperature was very hot. Most of the carbon dioxide remained in Earth's atmosphere until it was removed when

   a. ultraviolet radiation caused the gas to disassociate into carbon and oxygen.

   b. volcanic activity produced chemicals that neutralized the gas.

   c. liquid water formed on Earth's surface and dissolved the carbon dioxide causing it to react with other compounds to form limestone and other sediments.

   d. infrared radiation heated Earth's atmosphere and evaporated the carbon dioxide.

11. The form of electromagnetic radiation that the ozone layer prevents from reaching Earth's surface is

   a. X rays.

   b. visible light.

   c. ultraviolet radiation.

   d. infrared rays.

12. Ozone is actually a compound of the element

   a. hydrogen.

   b. oxygen.

   c. nitrogen.

   d. helium.

13. The temperature of Earth is much warmer and more uniform than it would be if Earth did not have an atmosphere. The gas mostly responsible for trapping heat in Earth's atmosphere is

   a. carbon dioxide.

   b. oxygen.

   c. helium.

   d. nitrogen.

14. The greenhouse effect best describes the process by which
    a. ultraviolet radiation is filtered from sunlight.
    b. infrared radiation is reflected off the top of Earth's atmosphere.
    c. carbon dioxide is dissolved by liquid water.
    d. heat is trapped by the atmosphere, thus increasing the temperature at Earth's surface.

15. Free oxygen in Earth's atmosphere was produced by
    a. chemical reactions taking place in the oceans.
    b. living organisms during photosynthesis.
    c. chemical reactions taking place on land.
    d. the decay of early plant life.

16. A major concern of scientists today is the overproduction of chemicals known as CFCs. These chemicals
    a. pollute ocean waters and destroy plant life.
    b. destroy the ozone layer in Earth's atmosphere.
    c. enrich the soil, producing an overabundance of plant life and causing the loss of soil nutrients.
    d. do nothing harmful. Scientists are misguided in their conclusions.

Using What You've Learned

1. Diagram the internal structure of Earth, identifying the four layers. Continue the diagram to include the five major regions of Earth's atmosphere. List the altitude of each region and identify where interplanetary space begins.
    a. In which layer of the atmosphere does weather occur?
    b. What is the hydrosphere?
    c. What is the lithosphere?

2. Answer the following questions:

    a. Why is the sky blue?

    b. Why does the sun sometimes appear larger at the horizon, just before sundown, than when it is overhead?

    c. How do halos form around the sun and the moon?

    d. What are mirages and how do they occur?

3. Visit the geology and earth science exhibits at a museum to learn more about Earth.

4. You discover a new satellite of Earth. You send radar waves to it and receive the radar exactly 2 seconds after the signal is transmitted. What is the distance to the satellite in kilometers?

5. Make a list of the scientific evidence that supports the existence of the supercontinent of Pangaea. For example: Identify land forms and rock deposits that are the same, but lie on two different continents. Identify animal and plant species (can be fossils) that exist on two continents separated by an ocean.

Lesson Review

Lesson 20
Planet Earth

Please Note: Use this matrix to guide your study and achieve the learning objectives of this lesson. It will also help you to view the video, which defines and demonstrates important concepts and skills as they relate to everyday life.

Learning Objective	Textbook	Telecourse Student Guide	Background Notes
1. Outline a four-stage history of Earth's formation.	pp. 346–347	Matching Activities: 1; Self-Test Questions: 1, 2.	
2. Describe a model of Earth's interior and explain how we have come to develop that model.	pp. 347–348	Matching Activities: 2, 3, 5; Self-Test Questions: 3, 4.	
3. Explain the current theory about how the magnetic field of Earth is produced.	p. 347	Matching Activities: 4; Self-Test Questions: 5, 6.	
4. Explain the theory of and evidence for plate tectonics and continental drift.	pp. 348, 350–351	Matching Activities: 6, 7, 8, 9, 12, 13, 14; Completion Activities: 1; Self-Test Questions: 7, 8.	
5. Describe and explain the processes that have changed Earth's atmosphere from its earliest formation to the present day.	pp. 348–349, 352	Matching Activities: 10, 11; Completion Activities: 2, 3; Self-Test Questions: 9, 10.	
6. Explain the cause and possible consequences of a depletion of Earth's ozone layer.	pp. 349, 352	Completion Activities: 2; Self-Test Questions: 11, 12.	
7. Explain the concept of global warming and describe the possible consequences.	pp. 348–349	Matching Activities: 10; Completion Activities: 2; Self-Test Questions: 13, 14.	
8. Describe the effects life on Earth may have on Earth's environment.	pp. 349, 352	Self-Test Questions: 15, 16.	

LESSON
21

The Moon and Mercury

Assignments

For the most effective study of this lesson, we suggest that you complete the assignments in the sequence listed below.

Before Viewing the Video Program

- Read the Overview and Learning Objectives for this lesson. Use the Learning Objectives to guide your reading, viewing, and thinking.
- Read Chapter 17, "The Earthlike Planets," sections 17-2 and 17-3, pages 352–362, in the *Horizons* textbook.
- Read Background Notes 21A, "Radar Astronomy," and 21B, "Synchronous Rotation," following Chapter 26 in this guide.
- Read the Viewing Notes in this lesson.

View "The Moon and Mercury" Video Program

After Viewing the Video Program

- Briefly note your answers to the questions listed at the end of the Viewing Notes.
- Review all reading assignments for this lesson, especially the Chapter 17 summary on page 378 in *Horizons*, Background Notes 21A and 21B for Unit IV, and the Viewing Notes in this lesson.
- Respond to the challenges presented by each Critical Inquiry in the reading assignment and write brief answers to review questions 5–8 at the end of Chapter 17 in *Horizons* to be certain you understand the text material.
- Complete the Review Activities in this guide to reinforce your understanding of important terms and concepts. Check your answers with the Answer Key and review when necessary.

- Take the Self-Test in this guide to measure your achievement of the Objectives. Check your answers with the Answer Key and review when necessary.

- Complete the Using What You've Learned activities and any other activities and projects assigned by your instructor.

Overview

In our exploration of the solar system, we now move to a close-up study of the object nearest to Earth, the moon. Of all the objects in the night sky, it is perhaps the most familiar. Throughout human history, its changing appearance has stimulated numerous mythologies. Even today, poetry and songs celebrate the moon as a romantic object, which may seem surprising in that our closest celestial neighbor is a cold and desolate world.

Galileo, using his small telescope, was the first to identify features on the lunar surface. He correctly labeled mountains and craters. He thought the smooth, dark areas were oceans of water, which he called *mare*. Later, larger telescopes revealed these "seas" to be plains of rock. Without an atmosphere and without evidence of running water on the surface, the moon is a lifeless, unchanging world.

The Apollo manned lunar missions confirmed that the moon and Earth were very different. Samples brought back by these missions showed that moon rocks lacked water and other low-temperature volatiles. Seismometers left on the surface of the moon revealed that the interior of the moon is quite cold and does not have a liquid core. Without a warm interior, the crust of the moon is locked in place, and plate tectonics do not occur.

For planetary scientists studying the moon, one of the major questions is, "Why are there such major differences between Earth and the moon?" The search for the answer to this question is like a detective story. Using observations, results of experiments, analysis of lunar samples, and computer modeling and simulations, planetary scientists have developed a model for the formation and evolution of the moon. The first part of this lesson examines that model and compares it with Earth's history. As you will learn, the moon is a world that followed an evolutionary path completely different from Earth's.

Although Earth and the moon are very different, another object in our solar system appears, at least at first glance, to be very similar to the moon. Mercury is a heavily cratered planet. The number and distribution of its craters are similar to the moon's. Mercury also has large lava-filled plains like the lunar maria. In fact, if one did not know the difference, photographs of the moon and Mercury could easily be confused. However, the gravitational influence Mercury had on the Mariner 10 spacecraft as it flew by the planet revealed that its interior was distinctly different from that of the moon.

In order for Mercury to influence the spacecraft in the way it did, Mercury had to be extremely dense; in fact, the overall density of Mercury is approximately equal to that of Earth. For Mercury to be so dense, it must have an iron core at least two-thirds the diameter of the planet. Furthermore, some of that core is liquid. This conclusion was drawn after Mariner 10 detected a weak magnetic field about the planet. It seems that Mercury is like the moon on the outside, but more like Earth on the inside.

Our study of the moon, Mercury, and the other planets helps us to understand the processes involved in the formation and evolution of Earth. By comparing similarities and differences of these other worlds, we can see how unique and special Earth truly is.

Learning Objectives

When you have completed all assignments for this lesson, you should be able to:

1. Identify the major features and characteristics of the surface of the moon and describe the various processes that formed them. *HORIZONS* TEXTBOOK PAGES 352–357; VIDEO PROGRAM.

2. List and describe the various geological findings (surface and interior) made by spacecraft missions (manned and unmanned) to the moon. *HORIZONS* TEXTBOOK PAGES 353–359; VIDEO PROGRAM.

3. Describe the various theories about the origin of the moon and state the advantages and disadvantages of each. *HORIZONS* TEXTBOOK PAGES 357–359; VIDEO PROGRAM.

4. Describe the surface features of Mercury as seen by the Mariner 10 spacecraft and explain their origin. *HORIZONS* TEXTBOOK PAGES 360–362; VIDEO PROGRAM.

5. Describe a model for the interior of Mercury. *HORIZONS* TEXTBOOK PAGES 360–362; VIDEO PROGRAM.

6. Describe the radar observations of Mercury and cite advantages of radar observations over traditional Earth-based optical observations. BACKGROUND NOTES 21A and 21B; VIDEO PROGRAM.

Viewing Notes

Two worlds, similar in outward appearance, but quite different inside, are the subject of this video program, "The Moon and Mercury." The video program begins by looking at the moon, the celestial object closest to Earth and one that has long been observed by humans. Edwin Krupp, an archaeoastronomer, describes the importance of the moon to ancient cultures. Other astronomers explain how Galileo's early telescopic observations of the moon changed Western civilization's perception of the moon. Next, the video program illustrates what we now know about the basic features of the moon—the heavily cratered highlands and the lava-filled lowlands, the maria—and describes how these features were formed. This segment also explains synchronous rotation and why the same side of the moon always faces Earth. It includes photographs taken by Soviet spacecraft in 1959 of the back of the moon and an explanation of why the two hemispheres have a different appearance.

The moon is unique in that it is the only extraterrestrial place that humans have personally explored and from which we have actual samples. NASA footage of the first Apollo landing on the moon in 1969 and of several lunar walks and explorations is accompanied by planetary astronomers' descriptions of what we have learned about the moon's structure and composition from seismometers placed on the moon and from lunar samples brought back by astronauts.

The video program's examination of the moon ends with descriptions of different hypotheses (fission, condensation, and capture) of how the moon was formed and discussion of the failure of most of them to account for specific facts about the moon. The large-impact hypothesis of moon formation is the one that is currently most accepted, and the co-developer of that theory, William K. Hartman, a senior scientist at the Planetary Institute, explains the theory. Computer simulation illustrates the large-impact hypothesis and how the moon might have been created when a Mars-sized object crashed into Earth, sending mantle material into space, where it ultimately condensed into the moon.

From the moon, the celestial object closest to Earth, the video program moves to Mercury, the planet closest to the sun. Unlike the moon, Mercury is difficult to observe through optical telescopes because it is small and so close to the sun. Planetary astronomers describe our early limited knowledge of Mercury. Most of what we know about Mercury has been learned fairly recently, from radar studies and from the Mariner 10 spacecraft, which photographed Mercury in 1974–1975. Those photographs, shown in this segment, reveal a cratered surface remarkably similar to the moon's, along with maria, also similar to those on the moon but lighter in color, and scarps (tall cliffs), which are not present on the moon. Planetary astronomers describe those features and their formation and compare Mercury's structure to the moon's. Mariner 10 also detected a magnetic field and large iron core; and astronomers explain how, on the inside, Mercury is similar to Earth. Radar studies, which have mapped the entire surface of Mercury, have provided much of our current knowledge of the planet's surface, its composition, and its possible patches of water ice at the poles. Radar images of Mercury illustrate the value of this tool for studying the inner solar system.

As you watch the video program, consider the following questions:

1. What features on the lunar surface were first identified by telescopes, and what actions produced those features?

2. Why has the moon not been reshaped by erosion and plate tectonics?

3. What evidence suggests that the moon is asymmetric? How does this asymmetry cause the near and far sides of the moon to appear so different?

4. How does the large-impact hypothesis of the formation of the moon account for its differences in chemical composition and density when compared with Earth?

5. Why is optical observation of Mercury so difficult?

6. What did we know about Mercury before any spacecraft mission?

7. What are the similarities and differences between the surfaces of Mercury and the moon?

8. What information have radar studies of Mercury provided that could not be obtained by Earth-based optical telescopes?

Review Activities

Matching

Match the terms listed below with the definitions that follow. Check your answers with the Answer Key and review any terms you missed.

_____ 1. ejecta

_____ 2. rays

_____ 3. vesicular basalt

_____ 4. mare

_____ 5. anorthosite

_____ 6. breccia

_____ 7. albedo

_____ 8. early heavy bombardment

_____ 9. fission hypothesis

_____ 10. condensation hypothesis

_____ 11. capture hypothesis

_____ 12. large-impact hypothesis

_____ 13. lobate scarp

_____ 14. radar astronomy

_____ 15. synchronous rotation

a. theory that the moon formed from debris ejected during a collision between Earth and a large planetesimal

b. a lunar lowland filled by successive flows of dark lava

c. theory that the moon and Earth formed when a rapidly rotating protoplanet split into two pieces

d. pulverized rock scattered by meteorite impacts on a planet

e. rock with trapped bubbles formed by solidified lava

f. intense cratering that occurred during the first 500 million years in the history of the solar system

g. rock composed of fragments of earlier rocks bonded together

h. white streamers radiating from some lunar craters

i. theory that Earth and the moon condensed simultaneously from the same cloud of material

j. rock of aluminum and calcium silicates found in the lunar highlands

k. theory that Earth's moon formed elsewhere in the solar nebula and was later captured by Earth

l. a curved cliff such as those found on Mercury

m. the ratio of the light reflected from an object divided by the light that hits the object; this ratio equals 0 for perfectly black and 1 for perfectly white

n. identical rotation and revolution periods

o. use of artificially produced radio signals reflected off objects and analyzed according to the time delays and Doppler shifts of the return signal

Completion

Fill each blank with the most appropriate term from the list for that paragraph. A term may be used once, more than once, or not at all. Check your answers with the Answer Key and review when necessary. If a question requires two or more answers in succession, they may be in any order, unless the question indicates otherwise.

1. Upon viewing the moon through a telescope, one can identify the moon's two types of terrain. The light areas, or _____, are heavily _____. The other type of terrain on the moon can be described as large, relatively _____ plains of _____ material. Galileo believed these plains were _____ and therefore named them _____. The relative ages of these

different terrains can be determined by counting the number of
_____ in each. Such a count indicates that the _____
are younger than the _____.

The exact composition of these two areas was determined when lunar
samples were returned by the manned _____ missions.
Samples from the maria were shown to contain _____, which
is a hardened form of _____. This finding confirms the theory
that the maria are the result of _____ that left enormous craters
that eventually filled with molten rock. The other lunar area, the
_____, consists of rocks known as _____. These
rocks are composed of silicates of _____ and _____.
Their color is _____, and their density is _____ than
that of the _____ found in the lowlands of the moon.

aluminum	craters	lower
anorthosite	darker	lowlands
Apollo	highlands	maria
basalt	impacts	seas
calcium	lava	smooth
cratered	lighter	vesicular basalt

2. Astronomers have considered several different hypotheses to explain
the origin of the moon. One hypothesis was that Earth, in its early
history, rotated much more _____ than it does today. Because
of this rotation, the moon _____ from Earth. To explain why
the moon has so little _____, this hypothesis said that this
major event occurred after Earth had _____. One of the
problems of this _____ hypothesis was that it could not
explain the differences in _____ between Earth and moon
rocks. Furthermore, the Earth-moon system should contain a great deal
more _____ than is currently observed.

A second hypothesis was that the moon and Earth formed from the
same cloud of material in the _____. The moon and Earth
_____ simultaneously as twin planets. This _____
hypothesis did not remain plausible because of the differences in the
_____ compositions and _____ of the moon and
Earth. Evidence of this difference is found in the fact that the moon is
very poor in _____. In other words, the moon has little
material that is easily _____, such as _____ bound

up in its rocks, while Earth is abundant in rocks that contain

_____.

A third hypothesis for the origin of the moon was that the moon formed _____ in the solar system and was _____ by Earth. This hypothesis, the _____ hypothesis, solves a few problems. It allows for the moon to have significant differences in _____ and _____. It also permits Earth to rotate as _____ as it does today. Even so, this hypothesis has its problems. There must be a very special set of circumstances for Earth to _____ the moon. The speed and _____ of the moon had to be such that when it approached Earth the moon would enter into _____ around Earth and not be shattered by _____ forces. Since the moon is so large, and with the numerous _____ that are possible, it is highly unlikely that the specific set of circumstances allowing a successful _____ would have been met.

At present, the most widely accepted hypothesis for the origin of the moon is called the _____ hypothesis. This hypothesis suggests that, during Earth's early history, an object, about as large as _____, _____ with Earth. Material from Earth's _____ was blown into space and eventually condensed to form the _____. This explanation could account for the present amount of _____ in the Earth-moon system. Furthermore, if Earth had already _____, it would explain why the moon has so little _____. The heat generated may very well have been enough for _____ and other _____ to have been removed from the material that formed the moon.

angular momentum	differentiated	orbit
broke away	differentiation	rapidly
capture	elsewhere	slowly
captured	evaporated	solar nebula
chemical	fission	tidal
collided	iron	trajectories
composition	large-impact	trajectory
condensation	mantle	volatiles
condensed	Mars	water
density	moon	

3. The surfaces of Mercury and the moon resemble each other. They appear to have a similar number of _____. Both objects also have _____. On Mercury, however, the _____ are not as noticeable because they are filled with a _____-colored _____ as opposed to a _____-colored _____ on the moon. Although many features on Mercury are similar to those on the moon, there is one noticeable difference. Mercury has large curved cliffs called _____, which do not have any parallels on the lunar surface. These cliffs are believed to be the result of _____ in the crust of Mercury as the _____ of the planet _____. Although not visible in the _____ photographs, radar observations of Mercury indicate the planet may have water ice at its _____. Signals sent from Earth in 1991 reflected off Mercury and returned as a _____ image. Analysis of this image indicated the presence of _____. It is believed that the ice is in the soil of Mercury and on the floors of deep _____. These regions are forever hidden from the intense heat of the sun's rays.

cooled	maria
craters	Mariner 10
darker	poles
interior	radar bright
lava	shrinkage
lighter	water ice
lobate scarps	

Self-Test

Select the one best answer for each question. Check your answers with the Answer Key and review when necessary.

1. Large, reasonably flat and dark areas that cover a good portion of the side of the moon facing Earth are referred to as lunar

   a. maria.

   b. craters.

   c. escarpments.

   d. mascons.

2. The craters on the lunar surface were produced by
   a. volcanic activity.

   b. subsurface erosion due to running water.

   c. impacts by asteroids and meteorites.

   d. excavation by aliens.

3. Lunar rocks found in the highlands are
   a. older than the rocks in the maria and composed mostly of basalt.

   b. younger than the rocks in the maria and composed of aluminum and calcium.

   c. rich in anorthocites, a light-colored, low-density rock.

   d. younger than the rocks in the maria and composed mostly of basalt.

4. The core of the moon
   a. is large and composed largely of iron.

   b. is small and contains little iron.

   c. mostly pulverized dust, like that found on the surface.

   d. was destroyed billions of years ago by meteorites in the early bombardment epoch.

5. The large-impact hypothesis proposes that the moon formed as a result of
   a. a large amount of material breaking off Earth due to Earth's rapid rotation.

   b. two large asteroids colliding, shattering, and condensing to form the moon.

   c. a young moon colliding with Earth, forming a ring system that later condensed to form the moon we see today.

   d. a Mars-sized object colliding with Earth and ejecting material from Earth's mantle, some of which fell back to Earth and the rest condensed to form the moon.

6. A problem with the fission hypothesis for the formation of the moon is the fact that

   a. the Earth-moon system has less angular momentum than would be expected from such a formation scenario.

   b. Earth and the moon have similar densities.

   c. it requires too many coincidental events.

   d. the moon is rich in volatiles.

7. The craters on Mercury are the result of

   a. impacts by asteroids and meteorites.

   b. volcanic activity.

   c. subsurface erosion.

   d. solar wind erosion.

8. Lobate scarps on Mercury were probably caused by

   a. meteorite impacts.

   b. shrinkage of the surface as the interior of the planet cooled.

   c. surface erosion by flooding.

   d. wind erosion.

9. Evidence for Mercury's iron core comes from

   a. large impact craters.

   b. unusually low overall density for a planet of its size.

   c. unusually high overall density for a planet of its size.

   d. large lava-filled maria.

10. The weak (although stronger than expected) magnetic field around Mercury detected by the Mariner 10 spacecraft indicates

   a. a solid core.

   b. solar flare activity.

   c. the interior of the planet is similar to that of the moon.

   d. the interior of the planet is similar to that of Earth.

11. Radar observations of Mercury were first used to determine the planet's

    a. rotation rate.

    b. chemical composition.

    c. period of revolution about the sun.

    d. inclination of its orbit.

12. Radar bright patches were observed at the north and south poles of Mercury. These patches suggest the presence of

    a. carbon dioxide gas.

    b. water ice.

    c. aluminum and calcium silicates in the soil.

    d. vesicular basalt in the planet's maria.

Using What You've Learned

1. Compare the similarities and differences between Mercury and the moon. Include not only the surface features but also their internal structures.

2. What problems would humans face if they tried to colonize Mercury?

3. For the next several lessons, you will be asked to locate the five naked-eye planets on the SC1 Equatorial Star Chart (provided for you by your instructor or available at your college bookstore), determine if they are visible at night, and then, using the constellation charts in Appendix B of *Horizons*, go outside and try to find the planets. (You can also review "The Motion of the Planets," on pages 24–25 in *Horizons*.)

The SC1 Equatorial Star Chart

The Equatorial Star Chart shows the position of the stars and constellations situated within 60° north and south of the celestial equator. The celestial equator, if you recall, divides the sky into northern and southern hemispheres, just as the equator on Earth divides our planet into two hemispheres. In other words, if you stood on Earth's equator and looked up, the celestial equator (represented by the straight horizontal line through the center of the chart) would be directly overhead.

You will also notice the curved line running through the star chart. If you look closely, the line runs through the twelve constellations of the zodiac. This line is the ecliptic, the path the sun appears to follow through the sky over the course of a year. The small numbers along the ecliptic are celestial longitudes that you can use to locate the planets. Remember that the orbital planes of the planets lie close to the ecliptic plane and therefore the planets themselves are located on or near the ecliptic.

Locating the Stars

You have probably already identified several familiar constellations and located some of the brightest stars. A quick way of locating the stars is to use their coordinates.

The coordinates of stars are given in terms of their right ascension and declination (similar to longitude and latitude on a globe, respectively). The declination determines the position of the stars north (+) or south (-) of the celestial equator. If you look along the vertical line through the center of the chart, the line is marked off in degrees. It starts at 0° on the celestial equator and goes to 60° north (+) and 60° south (-) of declination. Right ascension is the horizontal coordinate, measured in hours, minutes, and seconds. It starts at the same vertical line at 0 hours. Running along the bottom of the chart, the right ascension goes from 0 to 12 hours on the left half of the chart, and from 12 to 24 hours on the right half of the chart.

Using the following table, "The Brightest Stars," locate several stars by using their coordinates of right ascension and declination. Draw an imaginary horizontal line through the declination coordinate on the star chart and an imaginary vertical line through the right ascension coordinate. The two lines should meet at the position of the star.

If you want to compare the chart with the appearance of stars in the night sky, hold your chart overhead, with the top of the chart toward the north, while you are facing south. Those stars that are near 0 hours right ascension will be nearly overhead in the early evening during November; those at 6 hours in February; those at 12 hours in May; and those at 18 hours in August.

BRIGHTEST STARS

STAR	NAME	RIGHT ASCENSION		DECLINATION	
1. α CMa A	Sirius	06	44.2	-16	42
2. α Car	Canopus	06	23.5	-52	41
3. α Boo	Arcturus	14	14.8	+19	17
4. α Cen A	Rigil Kentaurus	14	38.4	-60	46
5. α Lyr	Vega	18	36.2	+38	46
6. α Aur	Capella	05	15.2	+45	59
7. β Ori A	Rigel	05	13.6	-08	13
8. α CMi A	Procyon	07	38.2	+05	17
9. α Ori	Betelgeuse	05	54.0	+07	24
10. α Eri	Achernar	01	37.0	-57	20
11. β Cen AB	Hadar	14	02.4	-60	16
12. α Aql	Altair	19	49.8	+08	49
13. α Tau A	Aldebaran	04	34.8	+16	28
14. α Vir	Spica	13	24.1	-11	03
15. α Sco A	Antares	16	28.2	-26	23
16. α PsA	Fomalhaut	22	56.5	-29	44
17. β Gem	Pollux	07	44.1	+28	05
18. α Cyg	Deneb	20	40.7	+45	12
19. β Cru	Beta Crucis	12	46.6	-59	35
20. α Leo A	Regulus	10	07.3	+12	04

The Greek letter α stands for "alpha," and β stands for "beta." According to the Bayer system of star identification, these symbols identify the brightest and second brightest stars, respectively, in the named constellation.

Locating the Planets

The following table provides the celestial longitudes of the sun and the five naked-eye planets at 10-day intervals. Select a day (near the present date) and locate the position of the sun on the ecliptic. (The celestial longitudes are the small numbers near the ecliptic on the chart.) Using the same date, locate Mercury. Using the "Observing the Sky" star charts in the back of *Horizons*, go outside (at sundown or just before sunrise) and try to observe the planet. If the planet is to the right of the sun's position on the ecliptic, it will be visible in the predawn sky, unless it is too close to the sun. When the planet is to the left of the sun, it will follow the sun in the sky and may be visible after sunset, if it is not too close to the sun. Of all the planets, Mercury is the most difficult to observe. Good luck hunting!

CELESTIAL LONGITUDES OF THE SUN AND PLANETS

YEAR	DATE	SUN	MERCURY	VENUS	MARS	JUPITER	SATURN
2003	Aug 18	145	172	144	338	148	99
2003	Aug 28	154	176	157	335	150	100
2003	Sep 7	164	172	169	333	152	101
2003	Sep 17	174	163	182	331	154	102
2003	Sep 27	183	166	192	330	156	102
2003	Oct 7	193	180	206	331	158	103
2003	Oct 17	203	197	219	333	160	103
2003	Oct 27	213	214	231	335	162	103
2003	Nov 6	223	230	244	339	164	103
2003	Nov 16	233	246	256	344	165	103
2003	Nov 26	243	260	269	349	167	102
2003	Dec 6	235	274	281	354	168	102
2003	Dec 16	264	282	293	0	168	101
2003	Dec 26	274	276	306	6	169	100
2004	Jan 5	284	266	318	12	169	99
2004	Jan 15	294	270	330	18	169	99
2004	Jan 25	304	281	342	24	168	98
2004	Feb 4	314	295	354	30	167	97
2004	Feb 14	325	311	6	37	166	97
2004	Feb 24	335	327	18	43	165	96
2004	Mar 5	345	345	29	49	164	96
2004	Mar 15	355	5	40	56	163	96
2004	Mar 25	5	22	50	62	161	97
2004	Apr 4	15	31	60	69	160	97
2004	Apr 14	24	29	69	75	160	97
2004	Apr 24	34	23	77	81	159	98
2004	May 4	44	22	83	88	159	99
2004	May 14	53	28	86	94	159	100

YEAR	DATE	SUN	MERCURY	VENUS	MARS	JUPITER	SATURN
2004	May 24	63	39	85	100	159	101
2004	Jun 3	73	55	81	107	160	102
2004	Jun 13	82	75	75	113	161	103
2004	Jun 23	92	97	71	119	162	105
2004	Jul 3	101	117	70	126	164	106
2004	Jul 13	111	134	73	132	165	107
2004	Jul 23	120	147	78	138	164	109
2004	Aug 2	130	156	85	145	169	110
2004	Aug 12	140	159	94	151	171	111
2004	Aug 22	149	153	103	157	173	112
2004	Sep 1	159	146	134	164	175	113
2004	Sep 11	169	151	124	170	177	114
2004	Sep 21	178	166	135	176	179	115
2004	Oct 1	188	184	147	183	181	116
2004	Oct 11	198	202	158	189	183	117
2004	Oct 21	208	218	170	196	185	117
2004	Oct 31	218	233	182	203	187	117
2004	Nov 10	228	247	194	209	189	117
2004	Nov 20	238	260	207	216	191	117
2004	Nov 30	248	267	219	223	193	117
2004	Dec 10	258	259	232	229	195	116
2004	Dec 20	268	250	244	236	196	116
2004	Dec 30	279	256	257	243	197	115
2005	Jan 9	289	268	269	250	198	114
2005	Jan 19	299	283	282	257	198	113
2005	Jan 29	309	298	294	264	199	113
2005	Feb 8	319	314	307	271	199	112
2005	Feb 18	329	332	319	278	198	111
2005	Feb 28	339	351	332	285	198	111
2005	Mar 10	349	7	344	292	197	110
2005	Mar 20	359	14	357	299	196	110
2005	Mar 30	9	9	9	307	194	110
2005	Apr 9	19	2	21	314	193	111
2005	Apr 19	29	4	34	321	192	111
2005	Apr 29	39	12	46	328	191	112
2005	May 9	48	25	58	336	190	112
2005	May 19	58	41	71	343	189	113
2005	May 29	68	61	83	350	189	114
2005	Jun 8	77	83	95	357	189	115
2005	Jun 18	87	103	107	4	189	116
2005	Jun 28	96	120	120	11	190	118
2005	Jul 8	106	132	132	17	190	119
2005	Jul 18	115	139	144	24	191	120
2005	Jul 28	125	139	156	30	193	121
2005	Aug 7	135	133	168	35	194	123

YEAR	DATE	SUN	MERCURY	VENUS	MARS	JUPITER	SATURN
2005	Aug 17	144	129	180	41	196	124
2005	Aug 27	154	136	192	45	198	125
2005	Sep 6	163	152	203	49	199	126
2005	Sep 16	173	171	215	51	201	127
2005	Sep 26	183	189	226	53	203	128
2005	Oct 6	193	206	238	53	206	129
2005	Oct 16	203	221	249	52	208	130
2005	Oct 26	213	234	259	49	210	131
2005	Nov 5	223	246	270	46	212	131
2005	Nov 15	233	251	279	42	214	131
2005	Nov 25	243	242	288	40	216	131
2005	Dec 5	253	235	295	38	218	131
2005	Dec 15	263	242	300	38	220	131
2005	Dec 25	273	255	301	39	222	130
2006	Jan 4	283	270	299	42	234	130
2006	Jan 14	294	286	294	45	225	129
2006	Jan 24	304	302	288	49	226	128
2006	Feb 3	314	319	286	53	227	127
2006	Feb 13	324	337	288	57	228	126
2006	Feb 23	334	352	293	62	229	126
2006	Mar 5	344	356	300	68	229	125
2006	Mar 15	354	349	308	73	229	125
2006	Mar 25	4	343	318	79	228	124
2006	Apr 4	14	347	328	84	227	124
2006	Apr 14	24	357	338	90	226	124
2006	Apr 24	34	11	349	96	225	124
2006	May 4	43	27	1	102	224	125
2006	May 14	53	47	12	107	223	126
2006	May 24	63	69	23	113	222	126
2006	Jun 3	72	89	35	119	221	127
2006	Jun 13	82	105	47	125	220	128
2006	Jun 23	91	116	59	132	219	129
2006	Jul 3	101	121	71	138	219	130
2006	Jul 13	110	119	83	144	219	132
2006	Jul 23	120	123	95	150	219	133
2006	Aug 2	129	112	107	156	220	134
2006	Aug 12	139	121	119	163	221	135
2006	Aug 22	149	138	131	169	222	137
2006	Sep 1	158	158	143	175	223	138
2006	Sep 11	168	177	156	182	225	139

YEAR	DATE	SUN	MERCURY	VENUS	MARS	JUPITER	SATURN
2006	Sep 21	178	193	168	188	227	140
2006	Oct 1	188	208	181	195	228	141
2006	Oct 11	197	221	193	201	230	142
2006	Oct 21	207	232	206	208	232	143
2006	Oct 31	217	235	218	215	235	144
2006	Nov 10	227	225	231	222	237	144
2006	Nov 20	237	219	243	229	239	145
2006	Nov 30	248	228	256	236	241	145
2006	Dec 10	258	243	269	243	243	145
2006	Dec 20	268	259	282	250	245	145
2006	Dec 30	278	274	294	257	247	144
2007	Jan 9	289	290	307	264	249	143
2007	Jan 19	299	307	319	272	251	143
2007	Jan 29	309	324	332	279	253	142
2007	Feb 8	319	337	344	286	255	141
2007	Feb 18	329	338	356	294	256	140
2007	Feb 28	339	328	9	302	257	140
2007	Mar 10	349	325	21	309	258	139
2007	Mar 20	359	332	33	317	259	138
2007	Mar 30	9	343	45	324	259	138
2007	Apr 9	19	358	57	332	259	138
2007	Apr 19	29	15	68	340	259	138
2007	Apr 29	39	35	80	347	258	138
2007	May 9	48	56	91	355	257	138
2007	May 19	58	76	102	3	256	138
2007	May 29	68	90	113	10	255	139
2007	Jun 8	77	99	122	18	254	140
2007	Jun 18	87	101	132	25	253	141
2007	Jun 28	96	96	140	32	252	141
2007	Jul 8	106	92	146	39	251	143
2007	Jul 18	115	95	151	46	250	144
2007	Jul 8	125	107	152	53	249	145
2007	Aug 7	134	126	150	60	249	146
2007	Aug 17	144	146	144	66	249	147
2007	Aug 27	154	165	139	72	250	149
2007	Sep 6	163	182	136	78	251	150
2007	Sep 16	173	196	137	83	252	151
2007	Sep 26	183	208	142	88	253	152
2007	Oct 6	193	217	148	93	255	153
2007	Oct 16	203	217	157	97	256	155

YEAR	DATE	SUN	MERCURY	VENUS	MARS	JUPITER	SATURN
2007	Oct 26	212	206	166	99	258	155
2007	Nov 5	222	204	176	101	260	156
2007	Nov 15	233	215	187	102	262	157
2007	Nov 25	243	231	198	101	264	157
2007	Dec 5	253	246	210	99	266	158
2007	Dec 15	263	262	222	95	269	158
2007	Dec 25	273	278	234	92	271	158
2008	Jan 4	283	294	246	88	273	158
2008	Jan 14	293	310	258	85	276	157
2008	Jan 24	304	322	270	84	278	157
2008	Feb 3	314	320	282	84	280	156
2008	Feb 13	324	309	295	85	282	155
2008	Feb 23	334	309	307	87	284	155
2008	Mar 4	344	317	319	90	286	154
2008	Mar 14	354	329	332	93	287	153
2008	Mar 24	4	344	344	97	289	152
2008	Apr 3	14	1	356	101	290	152
2008	Apr 13	24	21	9	106	291	151
2008	Apr 23	33	42	21	111	291	151
2008	May 3	43	61	33	116	292	151
2008	May 13	53	75	46	122	292	151
2008	May 23	62	81	58	127	291	151
2008	Jun 2	72	79	70	133	291	152
2008	Jun 12	82	74	83	138	290	152
2008	Jun 22	91	73	95	144	289	153
2008	Jul 2	101	79	107	150	288	154
2008	Jul 12	110	93	120	156	286	155
2008	Jul 22	120	112	132	162	285	156
2008	Aug 1	129	133	144	168	284	157
2008	Aug 11	139	152	156	175	283	158
2008	Aug 21	149	169	169	181	282	160
2008	Aug 31	158	183	181	187	282	161
2008	Sep 10	168	195	193	194	282	162
2008	Sep 20	178	202	205	200	282	163
2008	Sep 30	187	200	218	207	283	165
2008	Oct 10	197	189	230	214	284	166
2008	Oct 20	207	189	242	221	285	167
2008	Oct 30	217	202	254	228	286	168
2008	Nov 9	227	218	266	235	288	169
2008	Nov 19	237	234	278	242	289	170

YEAR	DATE	SUN	MERCURY	VENUS	MARS	JUPITER	SATURN
2008	Nov 29	247	250	290	249	291	170
2008	Dec 9	258	265	302	256	293	171
2008	Dec 19	263	281	313	264	296	171
2009	Dec 29	278	296	324	271	298	171
2009	Jan 8	288	306	335	279	300	171
2009	Jan 18	298	302	345	286	302	171
2009	Jan 28	308	292	355	294	305	170
2009	Feb 7	319	293	3	302	307	170
2009	Feb 17	329	303	10	310	310	169
2009	Feb 27	339	316	14	317	312	168
2009	Mar 9	349	331	15	325	314	168
2009	Mar 19	359	348	11	333	316	167
2009	Mar 29	9	7	5	341	318	166
2009	Apr 8	19	28	0	349	320	165
2009	Apr 18	28	46	359	357	322	165
2009	Apr 28	38	58	1	4	323	165
2009	May 8	48	61	6	12	324	164
2009	May 18	57	57	13	20	325	164
2009	May 28	67	52	21	27	326	164
2009	Jun 7	77	55	31	35	326	165
2009	Jun 17	86	63	41	42	326	165
2009	Jun 27	96	78	51	49	326	166
2009	Jul 7	105	98	62	56	326	166
2009	Jul 17	115	119	73	63	325	167
2009	Jul 27	124	139	85	70	324	168
2009	Aug 6	134	156	96	77	322	169
2009	Aug 16	143	170	108	184	321	170
2009	Aug 26	153	180	120	90	320	172
2009	Sep 5	163	185	132	96	319	173
2009	Sep 15	173	182	144	102	318	174
2009	Sep 25	182	172	156	108	317	175
2009	Oct 5	192	174	168	114	317	177
2009	Oct 15	202	188	181	119	317	178
2009	Oct 25	212	205	193	124	317	179
2009	Nov 4	222	222	205	128	317	180
2009	Nov 14	232	238	218	132	318	181
2009	Nov 24	242	253	231	135	319	182
2009	Dec 4	252	268	243	138	321	183
2009	Dec 14	262	282	256	139	322	183
2009	Dec 24	273	291	268	139	324	184

YEAR	DATE	SUN	MERCURY	VENUS	MARS	JUPITER	SATURN
2010	Jan 3	283	285	281	138	326	184
2010	Jan 13	293	275	293	135	328	184
2010	Jan 23	303	279	306	132	331	184
2010	Feb 2	313	289	319	128	333	184
2010	Feb 12	323	303	331	124	335	183
2010	Feb 22	333	318	344	121	338	183
2010	Mar 4	344	335	356	120	340	182
2010	Mar 14	354	354	8	120	343	181
2010	Mar 24	3	14	21	121	345	180
2010	Apr 3	13	31	33	123	347	180
2010	Apr 13	23	41	46	126	350	179
2010	Apr 23	33	41	58	129	352	178
2010	May 3	43	34	70	133	354	178
2010	May 13	52	32	82	137	356	177
2010	May 23	62	37	94	142	357	177
2010	Jun 2	72	48	106	147	359	177
2010	Jun 12	81	64	118	152	0	177
2010	Jun 22	91	84	129	158	1	178

Lesson Review

Lesson 21

The Moon and Mercury

Please Note: Use this matrix to guide your study and achieve the learning objectives of this lesson. It will also help you to view the video, which defines and demonstrates important concepts and skills as they relate to everyday life.

Learning Objective	Textbook	Telecourse Student Guide	Background Notes
1. Identify the major features and characteristics of the surface of the moon and describe the various processes that formed them.	pp. 352–357	Matching Activities: 1, 2, 4, 7, 8; Completion Activities: 1; Self-Test Questions: 1, 2.	
2. List and describe the various geological findings (surface and interior) made by spacecraft missions (manned and unmanned) to the moon.	pp. 353–359	Matching Activities: 3, 5, 6; Completion Activities: 1; Self-Test Questions: 3, 4.	
3. Describe the various theories about the origin of the moon and state the advantages and disadvantages of each.	pp. 357–359	Matching Activities: 9, 10, 11, 12; Completion Activities: 2; Self-Test Questions: 5, 6.	
4. Describe the surface features of Mercury as seen by the Mariner 10 spacecraft and explain their origin.	pp. 360–362	Matching Activities: 13; Completion Activities: 3; Self-Test Questions: 7, 8.	
5. Describe a model for the interior of Mercury.	pp. 360–362	Self-Test Questions: 9, 10.	
6. Describe the radar observations of Mercury and cite advantages of radar observations over traditional Earth-based optical observations.		Matching Activities: 14, 15; Self-Test Questions: 11, 12.	Background Notes 21A: Radar Astronomy; Background Notes 21B: Synchronous Rotation.

L E S S O N
22

Venus and Mars

Assignments

For the most effective study of this lesson, we suggest that you complete the assignments in the sequence listed below.

Before Viewing the Video Program

- Read the Overview and Learning Objectives for this lesson. Use the Learning Objectives to guide your reading, viewing, and thinking.

- Read Chapter 17, "The Earthlike Planets," sections 17-4 and 17-5, pages 362–378, in the *Horizons* textbook.

- Read Background Notes 22, "Plate Tectonics on Mars?" and review Background Notes 21A, "Radar Astronomy," following Lesson 26 in this guide.

- Read the Viewing Notes in this lesson.

View the "Venus and Mars" Video Program

After Viewing the Video Program

- Briefly note your answers to the questions listed at the end of the Viewing Notes.

- Review all reading assignments for this lesson, especially the Chapter 17 summary in *Horizons* and the Viewing Notes in this lesson.

- Respond to the challenges presented by each Critical Inquiry in the reading assignment and write brief answers to review questions 9–15 at the end of Chapter 17 in *Horizons* to be certain you understand the text material.

- Complete the Review Activities in this guide to reinforce your understanding of important terms and concepts. Check your answers with the Answer Key and review when necessary.

- Take the Self-Test in this guide to measure your achievement of the Objectives. Check your answers with the Answer Key and review when necessary.

- Complete the Using What You've Learned activities and any other activities and projects assigned by your instructor.

Overview

Venus and Mars have long fascinated both astronomers and casual observers. Venus, after the sun and moon, is the third brightest object in the night sky. Mars glows with a pale red color, distinct from all the other planets. From earliest times, these planets were included in lore and mythology throughout the world. The ancient Greeks and Romans linked Venus with beauty and fertility and Mars with war and power. With the invention of the telescope, observers of these two worlds could see the phases of Venus and the "canals" and seasons of Mars. Such observations increased astronomers' and the public's fascination with these planets.

Even seen through the largest telescopes, Venus appears as a bright, featureless disk against the dark sky. Early astronomers concluded that Venus must be covered with a layer of clouds and that these clouds were probably made of water vapor. It wasn't long before speculation led astronomers and others to believe that Venus was a tropical world with dense jungles and large dinosaurs roaming its surface. In contrast, Mars is the only planet whose surface can be seen in detail from Earth. Over one hundred years ago, astronomers observed that the Martian surface regularly changed in appearance; some concluded that those variations in color were the result of the growth of vegetation as Mars experienced its seasons. Scientists and nonscientists alike were intrigued by the thought that Mars and Venus might sustain life.

Although one of the main factors motivating planetary scientists to study Venus and Mars may have been the belief that life existed on these worlds, it is no longer a significant reason. The study of Venus and Mars is motivated by the need to understand the origin, evolution, and environments of these worlds in order to understand better the origin, evolution, and environment of Earth. It is comparative planetology, a theme that runs throughout these lessons on the solar system, that motivates scientists to study Venus and Mars.

Venus and Mars have been shaped by geologic and atmospheric forces similar to those on Earth. Venus and Mars are the two planets closest to Earth in the solar system. This fact enables planetary scientists to understand how only small differences in position in the solar system affect the evolution of a planet. For example, one would expect to see close similarities between Venus and Earth because of their similar size and density. The two planets, however, are quite different, and this lesson explains why they are different. In contrast, Mars is one-half the diameter of Earth and has an atmosphere only 1 percent as dense as

Earth's, and yet its surface features are more similar to those on Earth than are the features on Venus. Furthermore, some of the Martian landforms are larger than anything found on any of the other terrestrial planets. Venus and Mars seem to be planets in the extreme. They have both more and less of what Earth has.

Learning Objectives

When you have completed all assignments for this lesson, you should be able to:

1. Describe the Pioneer, Magellan, and Venera spacecraft findings about the physical and geological features on the surface of Venus. *HORIZONS* TEXTBOOK PAGES 362–369; BACKGROUND NOTES 21A; VIDEO PROGRAM.

2. Describe the characteristics of the atmosphere of Venus and explain the role of the "greenhouse effect" in the environment on the surface of the planet. *HORIZONS* TEXTBOOK PAGES 362–363, 365–368; VIDEO PROGRAM.

3. Compare and contrast Venus and Earth and explain why the two planets have such different environments. *HORIZONS* TEXTBOOK PAGES 362–368; VIDEO PROGRAM.

4. Describe and explain the origin of the geologic features found on the surface of Mars by spacecraft. *HORIZONS* TEXTBOOK PAGES 370–376; BACKGROUND NOTES 22; VIDEO PROGRAM.

5. Describe the characteristics of the Martian polar regions and compare them with Earth's. *HORIZONS* TEXTBOOK PAGES 374–375; VIDEO PROGRAM.

6. Describe the atmosphere of Mars and compare it with Earth's. *HORIZONS* TEXTBOOK PAGES 369–370; VIDEO PROGRAM.

7. Identify the two moons of Mars, describe their surface features, and explain what these features may reveal about the history of these two objects. *HORIZONS* TEXTBOOK PAGES 376–378.

Viewing Notes

This video program, through the use of computer-generated images from radar astronomy, provides close-up views of Earth's closest neighbors: Venus and Mars. A geophysicist and a planetary scientist introduce the two planets by noting some of the striking similarities between the two planets and Earth. The remainder of the video program then explores Venus and Mars, providing detailed information about each planet's geologic features and atmosphere.

The segments about Venus begin with optical telescope images of the cloud-shrouded planet while astronomers explain why Earth-based observation of Venus has been so difficult. Then, an astrophysicist describes how radar astronomy enabled planetary scientists to gather the first real data about Venus, including information on its rotation rate and direction. The video program next shows animation of how the Magellan spacecraft mapped Venus' surface with radar waves that penetrated the planet's cloud cover.

Data obtained from the Magellan program are converted by computer into three-dimensional aerial views of Venus. The video program shows computer-generated footage from NASA and the Jet Propulsion Laboratory of some of Venus' most dramatic features: Ishtar Terra, with Maxwell Montes, a mountain more than 12 kilometers high; craters; arachnoids; and coronae. (It is important to note that the vertical scale of these images has been exaggerated to enhance the differences in elevation.) Other animation shows how the entire planet would appear if it were stripped of its cloud cover. Several astronomers describe the geologic features of Venus, its interior structure, the apparent absence of plate tectonics, and specific similarities and differences between Venus and Earth.

The Venusian atmosphere is the next topic covered in the video program. Several scientists supply information on the planet's atmospheric pressure, the constituent gases, and surface temperature. A planetary scientist and planetary meteorologist explain how conditions on Venus have produced a runaway greenhouse effect and compare the conditions on Venus with those on Earth.

The last half of the video program explores Mars, the most Earthlike planet in our solar system. Astronomers describe the early ideas about "mythic" Mars that existed up until the 1930s: about areas thought to be vegetation that changed with seasons and about polar ice caps. Although these features were seen through optical telescopes, it was not until the Mariner and Viking missions to Mars that planetary scientists were able to understand the features they had observed on the planet.

The video program includes the first photographs from space of another world, taken by the Mariner 4 spacecraft in its flyby of Mars in 1965. These black-and-white pictures showed a cratered, barren world, quite different from that imagined by science-fiction fans. Later missions supplied data that were used to create computer-generated three-dimensional images of the Valles Marineris (the Grand Canyon of Mars) and of Olympus Mons (a shield volcano). These images illustrate the geologic characteristics of Mars.

The segment on Mars next examines the evidence for a history of water on the planet and possible explanations of why that water is no longer present. Spacecraft photos show the dry streambeds on the Martian surface, and computer animation illustrates water-filled streambeds drying up. Planetary scientists describe current thinking about the composition and behavior of the polar ice caps and about the large dust storms that produce terraces around the Martian poles.

The program concludes with Jeff Goldstein, an astrophysicist at the National Air and Space Museum, explaining that the laws of physics, although the same throughout our solar system, manifest themselves differently on these other worlds than on Earth. He notes that study of the physics on other worlds increases our understanding of the physical laws that govern phenomena on Earth.

As you watch the video program, consider the following questions:

1. What are the difficulties in observing the planet Venus?

2. How has radar astronomy allowed us to "see" the surface of Venus?

3. What explanation is given for the retrograde rotation of Venus?

4. Why doesn't Venus exhibit evidence of continental drift and plate tectonics?

5. What is the greenhouse effect and why does it cause such high temperatures on the surface of Venus?

6. What were the early Earth-based telescopic observations of Mars and what conclusions were drawn based on those observations?

7. What is a shield volcano and why are the ones on Mars so much higher than terrestrial volcanoes?

8. What is the evidence that suggests that Mars had large amounts of liquid water flowing on its surface some time in its past and where is the water on Mars now?

Review Activities

Matching

Match the terms listed below with the definitions that follow. Check your answer with the Answer Key and review any terms you missed.

_____ 1. Magellan

_____ 2. Maxwell Montes

_____ 3. Ishtar Terra

_____ 4. shield volcanoes

_____ 5. coronae

_____ 6. arachnoids

_____ 7. pancake domes

_____ 8. greenhouse effect

_____ 9. Olympus Mons

_____ 10. Valles Marineris

_____ 11. Tharsis region

_____ 12. carbon dioxide

_____ 13. Phobos

a. largest volcano on Mars

b. wide, low-profile volcanic cone produced by highly liquid lava

c. flat circular shapes approximately 65 kilometers across and 1 kilometer high that formed when viscous lava leaked through cracks on the surface of Venus

d. larger of the two moons of Mars

e. on Venus, geological patterns of circular and radial faults resembling spider webs

f. spacecraft that orbited Venus and mapped its surface using radar

g. highland region on Venus larger than the United States and containing one of the largest volcanoes on the planet

h. the process by which a carbon dioxide atmosphere traps heat and raises the temperature of a planetary surface

i. most abundant gas in the atmosphere of Mars and Venus

j. on Mars, area in which several large volcanoes are located, including the planet's largest

k. one of largest volcanoes on the surface of Venus

l. a 2,500-mile-long canyon on Mars

m. large circular features on the surface of Venus that appear to be slightly domed, apparently caused by rising plumes of molten rock below the crust

Completion

Fill each blank with the most appropriate term from the list for that paragraph. A term may be used once, more than once, or not at all. Check your answers with the Answer Key and review when necessary. If a question requires two or more answers in succession, they may be in any order, unless the question indicates otherwise.

1. In comparison with the atmosphere of Venus, Earth's atmosphere is many times _____ dense. The Venusian atmosphere is also composed mostly of the gas _____ and, as a result, the surface temperature is hot enough to melt lead. The high surface temperature is produced by a phenomenon called the _____. Because the temperatures on Venus are so high, this phenomenon is often referred to as a _____.

Although the atmosphere of Venus is _____ and acts as a _____, it still allows the sun's _____ to reach the surface of the planet. This energy is absorbed by the _____ and reradiated away in the form of _____ energy. Because of the content of the atmosphere, much of the energy does not escape into space, but is trapped by the _____ in the atmosphere. As a result, the surface of the planet _____ in temperature to extremely _____ levels.

blanket	lead
carbon dioxide	less
decreases	low
escape	more
extreme	radiation
greenhouse effect	runaway greenhouse effect
high	surface
increases	thick
infrared	trapped

2. The major constituents of Earth's atmosphere are _____ and _____. However, Earth's atmosphere also has _____, and, as a result, Earth also experiences a _____, which increases the surface temperature. The heating that occurs on Earth, however, is much _____ than it is on _____. The principal reason for this difference is because Earth has _____ on its surface.

The early atmospheres of both Venus and Earth probably contained abundant quantities of the gases _____ and _____. However, because Earth is _____ from the sun, its early temperature was _____, which allowed the _____ in the atmosphere to condense into _____. The atmospheric _____ began to _____ into the _____, forming mineral deposits and carbonate rocks. In other words, much of the _____ that once was in the atmosphere of Earth is now in the shells of shellfish and at the bottom of the oceans in the form of _____ deposits.

carbon dioxide

dissolve

extreme

farther

greenhouse effect

increase

less

limestone

liquid water

lower

Mars

nearer

nitrogen

oceans

oxygen

Venus

water vapor

3. Our earliest detailed knowledge of the surface of Mars comes from the images sent back to us from the ＿＿＿＿＿＿ and ＿＿＿＿＿＿ spacecraft that explored Mars in the ＿＿＿＿＿＿. These spacecraft revealed Mars to be a geologically interesting place. The terrain in the southern hemisphere is heavily ＿＿＿＿＿＿, which gives the landscape the appearance of the ＿＿＿＿＿＿. The northern hemisphere has fewer ＿＿＿＿＿＿, and they are less ＿＿＿＿＿＿ than those in the southern hemisphere and therefore ＿＿＿＿＿＿. Much of the terrain in the northern hemisphere is due to extensive ＿＿＿＿＿＿ that has buried the original surface.

The volcanoes on Mars that have produced extensive ＿＿＿＿＿＿ are of the ＿＿＿＿＿＿ type, similar to the volcanoes of ＿＿＿＿＿＿. However, the volcanoes on Mars are larger than the volcanoes on Earth. The size difference in volcanoes results from the weaker gravity on Mars and from a ＿＿＿＿＿＿ Martian crust. This characteristic of the Martian crust is due to the fact that Mars is ＿＿＿＿＿＿ than Earth and therefore ＿＿＿＿＿＿ more quickly, allowing for a ＿＿＿＿＿＿ crust to develop. This crust enabled the volcanoes to ＿＿＿＿＿＿ in size until they are much ＿＿＿＿＿＿.

However interesting the observed geologic activity on Mars was, planetary scientists were especially interested to find ＿＿＿＿＿＿. These features indicate that, at some time in the Martian past, there must have been ＿＿＿＿＿＿ on the surface. In fact, some observed features, such as ＿＿＿＿＿＿, indicate that there may have been major ＿＿＿＿＿＿ on the Martian landscape. Today, however, Mars has no ＿＿＿＿＿＿ on its surface. However, there may be large quantities of _____ under the surface and at the Martian _____. The present lack of _____ on the surface demonstrates that the Martian climate has changed. The atmosphere of Mars is _____ than it was in the past and can no longer maintain a high enough _____ to keep _____ in its _____ form. The change in atmosphere probably results from the fact that Mars is _____ than Earth and its _____ cannot hold onto the gases to maintain a more _____ atmosphere.

Alaska	larger	pressure
cooled	lava flows	runoff channels
cratered	liquid	shield
craters	liquid water	smaller
dense	Mariner	thicker
eroded	Moon	thinner
flooding	1970s	Viking
gravity	1980s	volcanic activity
grow	outflow channels	water
Hawaii	poles	younger
ice		

Self-Test

Select the one best answer for each question. Check your answers with the Answer Key and review when necessary.

1. Radar mapping of Venus by spacecraft discovered that roughly 75 percent of the surface of Venus consists of
   a. volcanoes.
   b. fault zones.
   c. craters.
   d. basaltic lowlands.

2. Spacecraft images (radar) of Venus offer little evidence for
   a. volcanic activity.
   b. major plate motion.
   c. the existence of mountains on the surface.
   d. continent-sized land masses.

3. Analysis of the surface by spacecraft that have landed on Venus indicates that the surface material consists of
   a. basalt.
   b. water.
   c. liquid magma.
   d. silicate carbonates.

4. The Magellan and Venera spacecraft located similar features, called coronae, on the surface of Venus. These features are
   a. large craters produced by asteroid impacts.
   b. volcanic cones.
   c. circular features probably caused by rising columns of magma under the crust.
   d. eroded shield volcanoes.

5. The major component of the atmosphere on Venus is
   a. nitrogen.
   b. oxygen.
   c. sulfuric acid.
   d. carbon dioxide.

6. The high temperatures on the surface of Venus are caused by
   a. the planet's slow rotation rate.
   b. volcanic activity on the surface.
   c. a (severe) "runaway" greenhouse effect.
   d. reflective properties of the atmosphere.

7. The atmosphere of Venus contains much more carbon dioxide than the atmosphere of Earth. This difference is probably the result of
   a. the presence of more carbon dioxide in the primeval atmosphere of Venus.
   b. the slightly warmer temperatures on Venus that prevented the accumulation of great amounts of liquid water, thus preventing the dissolution of carbon dioxide.
   c. most of the carbon dioxide in Earth's atmosphere escaping into space.
   d. carbon dioxide still being produced by volcanoes on Venus.

8. Earth's crust is broken into large plates that slide over Earth's mantle. On Venus,
   a. similar plate activity has been observed.
   b. the plates are much smaller and move more quickly.
   c. there is no sign of plate tectonics.
   d. there is little evidence that plate tectonics played a major role in shaping the surface of the planet.

9. Evidence of liquid water existing in the Martian past is found by the discovery of

    a. meandering drainage valleys and elongated flow features exposing ancient craters.

    b. oceans of liquid water currently on the surface.

    c. a dense atmosphere.

    d. volcanic activity.

10. Evidence of extensive volcanic activity on Mars is found in the

    a. highly cratered southern hemisphere of Mars.

    b. Tharsis region in the northern hemisphere where there are numerous faults and lava flows.

    c. polar ice caps where spectrum analysis shows evidence of basalt rocks.

    d. southern hemisphere where volcanoes such as Olympus Mons are located.

11. The polar caps of Mars consist of

    a. water ice.

    b. solid methane and carbon dioxide.

    c. solid nitrogen ice and liquid water.

    d. carbon dioxide and frozen water.

12. In the polar ice caps on Mars, water

    a. melts in the spring.

    b. never melts.

    c. evaporates in the summer.

    d. is not present.

13. The atmosphere of Mars consists primarily of

    a. carbon dioxide.

    b. oxygen.

    c. nitrogen.

    d. water vapor.

14. The density of the atmosphere on Mars is about
    a. equal to Earth's atmosphere.

    b. 50 percent of Earth's atmosphere.

    c. 10 percent of Earth's atmosphere.

    d. 1 percent of Earth's atmosphere.

15. The moons of Mars are
    a. Phobos and Deimos.

    b. Io and Ganymede.

    c. Hellas and Ishtar Terra.

    d. Maxwell Montes and Elysium.

16. The moons of Mars can be briefly described as
    a. smooth and nearly spherical in shape.

    b. being lightly cratered.

    c. irregular in shape with numerous surface craters.

    d. nearly identical in surface features.

Using What You've Learned

1. Locate the planets Venus and Mars on the SC1 Star Chart. Use the longitudes given in the table in Lesson 21. Choose a date near the current one and locate the sun, then determine if Venus or Mars can be seen in the nighttime sky. If you can see one of the planets, go out and observe it.

2. Compare the characteristics of the atmosphere and surface features of Venus with those of Earth. Do you feel that, in some respects, Earth is becoming more like Venus? Give examples.

3. Develop a diagram, including Earth, the sun, and Venus, that shows the cycle of the phases of Venus as viewed from Earth. If Venus is currently visible, use a pair of binoculars or a small telescope to observe the planet and determine where on your diagram Venus must be in order for you to observe its current phase.

4. Try to observe Mars with a small telescope. Draw the surface features that you can see. Try to identify these features on a detailed map of Mars.

5. Read a science-fiction novel about the planet Mars. For example, read *Princess of Mars* by Edgar Rice Burroughs (1912). Compare how Mars is described in this book with what we know about it today.

6. Research the dimensions of Olympus Mons and Mauna Loa, and of Valles Marineris and the Grand Canyon. Construct a scale model of these surface features and compare their sizes.

Lesson Review
Lesson 22
Venus and Mars

Please Note: Use this matrix to guide your study and achieve the learning objectives of this lesson. It will also help you to view the video, which defines and demonstrates important concepts and skills as they relate to everyday life.

Learning Objective	Textbook	Telecourse Student Guide	Background Notes
1. Describe the Pioneer, Magellan, and Venera spacecraft findings about the physical and geological features on the surface of Venus.	pp. 362–369	Matching Activities: 1, 2, 3, 4, 5, 6; Self-Test Questions: 1, 2, 3, 4.	Background Notes 21A: Radar Astronomy
2. Describe the characteristics of the atmosphere of Venus and explain the role of the "greenhouse effect" in the environment on the surface of the planet.	pp. 362–363, 365–368	Matching Activities: 8; Completion Activities: 1, 2; Self-Test Questions: 5, 6.	
3. Compare and contrast Venus and Earth and explain why the two planets have such different environments.	pp. 362–368	Matching Activities: 5, 6, 7; Completion Activities: 2; Self-Test Questions: 7, 8.	
4. Describe and explain the origin of the geologic features found on the surface of Mars by spacecraft.	pp. 370–376	Matching Activities: 9, 10, 11; Completion Activities: 3; Self-Test Questions: 9, 10.	Background Notes 22: Plate Tectonics on Mars?
5. Describe the characteristics of the Martian polar regions and compare them with Earth's.	pp. 374–375, 363	Completion Activities: 3; Self-Test Questions: 11, 12.	
6. Describe the atmosphere of Mars and compare it with Earth's.	pp. 369–370	Matching Activities: 12; Self-Test Questions: 13, 14.	

Learning Objective	Textbook	Telecourse Student Guide	Background Notes
7. Identify the two moons of Mars, describe their surface features, and explain what these features may reveal about the history of these two objects.	pp. 376–378	Matching Activities: 13; Self-Test Questions: 15, 16.	

Jupiter and Saturn

Assignments

For the most effective study of this lesson, we suggest that you complete the assignments in the sequence listed below.

Before Viewing the Video Program

- Read the Overview and Learning Objectives for this lesson. Use the Learning Objectives to guide your reading, viewing, and thinking.
- Read Chapter 18, "Worlds of the Outer Solar System," sections 18-1 and 18-2, pages 381–397, in the *Horizons* textbook.
- Read Background Notes 23, "The Galileo Probe into the Atmosphere of Jupiter," following Lesson 26 in this guide.
- Read the Viewing Notes in this lesson.

View the "Jupiter and Saturn" Video Program

After Viewing the Video Program

- Briefly note your answers to questions listed at the end of the Viewing Notes.
- Review all reading assignments for this lesson, especially the first six paragraphs of the Chapter 18 summary on pages 408–409 in *Horizons* and the Viewing Notes in this lesson.
- Respond to the challenges presented by each Critical Inquiry in the reading assignment and write brief answers to review questions 1–8 and 10 at the end of Chapter 18 in *Horizons* to be certain you understand the text material.
- Complete the Review Activities in this guide to reinforce your understanding of important terms and concepts. Check your answers with the Answer Key and review when necessary.
- Take the Self-Test in this guide to measure your achievement of the Objectives. Check your answers with the Answer Key and review when necessary.
- Complete the Using What You've Learned activities and any other activities and projects assigned by your instructor.

Overview

This lesson begins the tour of the cold regions of the outer solar systems, and the first worlds visited are Jupiter and Saturn, the two largest planets in the solar system. Both are gas/liquid giants and both have a family of orbiting moons and rings encircling them. Jupiter is five times the Earth-sun distance, and Saturn is ten times the Earth-sun distance. These planets, at such great distances from the sun, are not made of the familiar materials found on Earth. Hydrogen and helium, so rare on Earth, are the major components of these worlds. Not only are these elements abundant on these planets, they are in forms that are difficult, if not impossible, to duplicate in Earth laboratories. The environment of the outer planets is extremely unlike that of the inner planets. The temperatures at which the outer planets formed were much lower and the results are strikingly different. Even the moons of these worlds are also unusual. Instead of being made of the same materials as the planets they orbit, such as is the case with our moon, most of these satellites are composed of more ice than rock. One of the moons has an atmosphere. In fact, at increasingly greater distances from the sun, the moons of the outer planets tend to contain ever-higher percentages of ice.

The most massive object in the solar system, except for the sun, is Jupiter. Visible from Earth as the fourth-brightest object in the sky, Jupiter has retained its royal position in mythology and in current planetary research. The planet is an environment of extremes that provides astronomers with an opportunity to observe phenomena that cannot be duplicated on Earth.

Jupiter is surrounded by a satellite system. In orbit around the planet are more than a dozen moons, four of which have been photographed by the Voyager spacecraft. Those images reveal surfaces and environments that have not been observed elsewhere in the solar system: active volcanoes, smooth surfaces that have cracked and refrozen, cratered terrain, grooves, and multiple ringed craters. With its many unique features, the Jupiter system provides many challenges for planetary scientists.

Saturn, the other giant of the solar system, is almost as large as Jupiter. Unlike Jupiter, Saturn has an extensive system of rings. Although astronomers have known for quite some time that the rings were not solid, but made of small, icy particles, the arrangement of those particles was totally unexpected. The Voyager images showed such an intricate arrangement of ring particles that scientists had to explore new applications of Newton's law of universal gravitation.

Like Jupiter, Saturn has a satellite system in orbit around it. However, these moons are different from the moons of Jupiter and those in the inner solar system. They contain more ice than rock. One moon even has an atmosphere that is denser than Earth's. As astronomers learn more about Jupiter, Saturn, and their moons, they have realize that additional factors must be considered when explaining the nature and formation of the diverse objects in the solar system. Not only must the material of which the planets and moons are composed be accounted for, but their distance from the sun also needs to be included in the picture.

What astronomers learn from studying these immense, cold worlds is one more contribution to our understanding of the content, origin, and evolution of the solar system.

Learning Objectives

When you have completed all assignments for this lesson, you should be able to:

1. Describe the size, shape, rotation, and orbital characteristics of Jupiter and Saturn and describe their appearance as seen through a telescope. *HORIZONS* TEXTBOOK PAGES 381–388, 391–392, AND 394–395; VIDEO PROGRAM.

2. Recount the discoveries made by the Voyager and Galileo spacecraft at Jupiter and the Voyager spacecraft at Saturn. *HORIZONS* TEXTBOOK PAGES 381–397.

3. Identify the Galilean satellites in orbit around Jupiter, provide details of their physical characteristics, and explain the origin of the identified features. *HORIZONS* TEXTBOOK PAGES 388–391; VIDEO PROGRAM.

4. Compare and contrast the ring systems of Jupiter and Saturn and explain the origin of these ring systems. *HORIZONS* TEXTBOOK PAGES 386–388, 392, AND 394–395; VIDEO PROGRAM.

5. Describe the current models of the atmosphere and interior of Jupiter and Saturn and explain the similarities and differences. *HORIZONS* TEXTBOOK PAGES 382–386 AND 391–393.

6. Identify the major moons in orbit around Saturn, describe their physical characteristics, and explain the origin of these features. *HORIZONS* TEXTBOOK PAGES 392–397; VIDEO PROGRAM.

7. Describe in detail the characteristics of the extensive ring system around Saturn and explain its structure. *HORIZONS* TEXTBOOK PAGES 394–395; VIDEO PROGRAM.

8. Compare and contrast the characteristics and physical appearance of the large storms (spots) on Jupiter and Saturn. *HORIZONS* TEXTBOOK PAGES 385–386 AND 391; VIDEO PROGRAM.

Viewing Notes

The two giants of the outer solar system—Jupiter and Saturn—are featured in this video program.

The first half of the program covers Jupiter and includes images of that planet taken on the Voyager missions, which explored the outer solar system and provided astronomers with vast quantities of information. In the program, astronomers describe what is currently known about Jupiter's features, including its unusual liquid composition of principally hydrogen and helium under tremendous pressure and its colorful atmosphere, full of cloud patterns and large areas of violent storms, and marked with the Great Red Spot. Images of Jupiter taken by spacecraft illustrate many of the features described in the program. This segment also includes information on Jupiter's rapid rotation and magnetic field.

Jupiter, along with the other Jovian planets, has both moons and rings. The Voyager explorations have provided excellent images of those objects, and astronomers explain what they have learned about moons Io, Europa, Ganymede, and Callisto and about the small particles that form the rings.

Saturn is featured in the second half of the video program. Astronomers point out the many similarities between the atmospheres and interiors of Saturn and Jupiter as well as noting some significant differences. Particular attention is focused on Saturn's spectacular concentric ring system, including descriptions of the system's structure and theories about the origin of the rings.

The segment on Saturn also includes information on Saturn's moons, including Titan with its cold atmosphere (rich in nitrogen but also having methane) and what has been learned from atmospheric studies and radar observations of Titan. Astronomers briefly describe Enceladus and Mimas, two of Saturn's other moons.

As you watch the video program, consider the following questions.

1. What are the primary components of Jupiter and Saturn? What are the gases?

2. How does the rapid rotation of Jupiter affect its cloud patterns?

3. What is the Great Red Spot?

4. What evidence, observed from Earth, revealed that Jupiter had a magnetic field? How is the magnetic field produced?

5. How do the characteristics of the Galilean satellites depend on their distance from Jupiter?

6. How do scientists explain the volcanic activity on Io?

7. What is the source of particles for Jupiter's ring?

8. How is the physical appearance of Saturn's clouds different from that of Jupiter's?

9. What is the composition and structure of Saturn's rings?

10. How do astronomers explain the intricate structure of Saturn's rings?

11. What are the similarities and differences between the moons of Jupiter and Saturn?

Review Activities

Matching

Match the terms listed below with the definitions that follow. Check your answers with the Answer Key and review any terms you missed.

_____ 1. magnetosphere   _____ 5. tidal heating

_____ 2. belt-zone circulation   _____ 6. oblateness

_____ 3. Roche limit   _____ 7. shepherd satellite

_____ 4. grooved terrain   _____ 8. forward scattering

a. heating of a planet or satellite because of friction caused by tidal forces

b. minimum distance between a planet and a satellite that holds itself together by its own gravity

c. volume of space around a planet within which the motion of charged particles is dominated by the planetary magnetic field rather than by the solar wind

d. atmospheric behavior typical of Jovian planets

e. flattening of a spherical body

f. regions of surface of Ganymede produced by repeated fracture and refreezing of the icy crust

g. object that, by its gravitational field, confines particles to a planetary ring

h. optical property of finely divided particles to preferentially direct light in the original direction of the light's travel

Completion

Fill each blank with the most appropriate term from the list for that paragraph. A term may be used once, more than once, or not at all. Check your answers with the Answer Key and review when necessary. If a question requires two or more answers in succession, they may be in any order, unless the question indicates otherwise.

1. The four largest moons of Jupiter were discovered by _____ and are referred to as the _____ satellites. The moon nearest Jupiter is _____. It is quite different from the other moons because is has virtually no _____, unlike the abundant number found on our own moon. The fact that this satellite has no

_____ indicates that its surface is very young. Images taken by the _____ spacecraft revealed the reason. This moon is covered with active _____. The cause of this activity appears to be this moon's _____ orbit. The gravitational influence of Jupiter varies as the moon's distance from the planet varies. The moon is flexed by _____, which generate tremendous amounts of _____ by internal friction.

The other three moons, or satellites, are equally as interesting. The second one out is _____. It also has no _____ on its surface, indicating that the surface is also very _____ and active. Its density is higher than those of the outer two moons, indicating a proportionally larger core. This core is believed to be covered by a _____ mantle and a thick, _____ crust. The largest of Jupiter's moons is _____, the third of the _____ satellites. This moon is _____ and is also marked by _____, which appears to be breaks in the crust. These features formed _____ the period of cratering and are therefore relatively _____. The last of the four large satellites of Jupiter is _____. It is the most _____ of all the other moons and therefore has the _____ surface. Its density is less than half that of Earth's density. This low density is best explained by a composition that is a fifty-fifty mixture of _____ and _____.

after	Ganymede	oldest
before	gas	rock
Callisto	grooved terrain	rocky
cratered	heat	tidal forces
craters	ice	Titan
elliptical	icy	volcanoes
Europa	Io	Voyager
Galilean	liquid water	young
Galileo		

2. The most impressive feature of the planet Saturn is its spectacular ring system. The first to observe the rings was _____, although he did not know what he was actually viewing. From Earth, the rings were discovered to contain billions of small _____ and a few _____. The largest break in the rings is called the _____ division. Details of the rings were revealed in images obtained by the _____ spacecraft. These images showed that the rings are very complex, consisting of thousands of narrow _____. In addition, new rings were discovered. One ring in particular was very twisted and braided. The structure of this ring is caused by _____ small _____. As a result of the _____ influences of these objects, the ring particles are kept in a very narrow band. These objects are called _____ satellites. Other gaps in the rings are caused by gravitational _____ between the ring particles and _____. For example, the largest break in the rings is the result of a tug of war between the ring particles and the moon _____. Remember, however, not all the breaks in the rings can be explained by this kind of gravitational tug of war, especially the hundreds of _____.

Cassini	moons	shepherding
Galileo	particles	three
gaps	resonances	two
gravitational	ringlets	Voyager
Mimas	satellites (moons)	

Self-Test

Select the one best answer for each question. Check your answers with the Answer Key and review when necessary.

1. The largest planet in the solar system is
   a. Jupiter.
   b. Saturn.
   c. Uranus.
   d. Neptune.

2. The oblateness of Jupiter and Saturn is the result of their
   a. ring system.
   b. rapid rotation.
   c. numerous satellites.
   d. gravitational resonances.

3. As viewed through a telescope, Jupiter appears as a
   a. spherical and heavily cratered object.
   b. bland and colorless disk.
   c. disk composed of various colored belts and bands.
   d. highly elliptical and bland disk.

4. Saturn, as viewed through a telescope, appears as a
   a. spherical and heavily cratered object.
   b. bland and nearly featureless disk.
   c. disk composed of brightly colored belts and bands.
   d. highly elliptical and bright blue disk.

5. Spacecraft photos revealed active volcanoes on the surface of Jupiter's moon
   a. Callisto.
   b. Titan.
   c. Ganymede.
   d. Io.

6. The Voyager 1 spacecraft discovered a ring system around
   a. Io.
   b. Saturn.
   c. Jupiter.
   d. Ganymede.

7. The grooved terrain on Ganymede is believed to be
   a. volcanic activity on the surface.
   b. systems of faults in the brittle crust.
   c. asteroid impacts.
   d. cometary impacts.

8. The surface of Io is covered with
   a. ice.
   b. sulfur ash.
   c. nitrogen.
   d. methane.

9. The ring particles around Jupiter are very small, as evidenced by the observation that light
   a. produces a rainbow of colors.
   b. reflects off the particles in the backward direction.
   c. scatters in the forward direction.
   d. is absorbed by the ring particles.

10. The pressure of sunlight and magnetic forces around Jupiter would cause the rings around Jupiter to dissipate relatively quickly. Therefore, the rings of Jupiter are resupplied by
    a. dust particles blasted into space as the result of meteorite impacts on the inner moons.
    b. cometary disintegration in the vicinity of Jupiter.
    c. small moons coming within the Roche limit and shattering.
    d. volcanic ash ejected into space by the volcanoes on Io.

11. The icy rings of Saturn are most likely not composed of particles left over from the formation of the planet but rather

    a. remnants of a moon that came within the planet's Roche limit.

    b. debris from collisions between the heads of comets and icy moons.

    c. captured micrometeorites.

    d. material captured as comets pass within the planet's Roche limit.

12. The core of Jupiter is composed of

    a. rock and ice.

    b. liquid and gaseous hydrogen.

    c. liquid helium.

    d. heavy elements such as iron, nickel, and silicon.

13. The belts and bands of Jupiter are the result of atmospheric features that have been

    a. brought up by convection currents from deep within Jupiter's interior.

    b. refracted through the hydrogen and helium atmosphere.

    c. distorted by cometary impacts.

    d. stretched and elongated as a result of Jupiter's rapid rotation.

14. The atmosphere of Titan is composed primarily of

    a. oxygen.

    b. nitrogen.

    c. methane.

    d. nothing; there is no atmosphere on Titan.

15. Some regions of Saturn's moon Enceladus are a thousand times less cratered than other regions. These lightly cratered regions

    a. missed the cratering epoch in the evolution of the solar system.

    b. probably have craters so small they were not observed by Voyager.

    c. are relatively young compared to the other more heavily cratered regions.

    d. are very old compared to the other more heavily cratered regions.

16. The Cassini division is a large gap in the rings of Saturn. This gap is most likely produced by

    a. the reflection of sunlight off dark material.

    b. resonances between ring particles and the moon Mimas.

    c. shepherding satellites.

    d. the absorption of sunlight by dark material.

17. Two small _____ interact gravitationally with wandering ring particles to keep Saturn's irregularly twisted F ring confined.

    a. gaps

    b. shepherd satellites

    c. artificial satellites

    d. comets

18. The extended life of Jupiter's Great Red Spot is probably because the Great Red Spot

    a. originates from the core of Jupiter.

    b. is related to the position of the moon Io.

    c. lies continually over a liquid "surface."

    d. generates its own thermal energy.

19. The belts and zones in the atmosphere of Saturn are less visible than those on Jupiter because

    a. of thermal radiation.

    b. of extreme temperatures, both high and low.

    c. of the gravitational pull of Titan.

    d. they occur deeper in the cold atmosphere below a layer of methane haze.

20. Examination of photographs of Europa taken by the Galileo spacecraft in orbit around Jupiter support the conclusion that

    a. liquid water exists beneath the icy crust.

    b. the core of Europa is rocky.

    c. volcanoes are currently erupting.

    d. life exists beneath the surface.

Using What You've Learned

1. Refer to an astronomy magazine that provides observing information for Jupiter and the Galilean satellites of Jupiter. With a small telescope or good binoculars, identify the four Galilean satellites and sketch their positions. Keep observing these moons for a period of several hours over the course of an evening. Which moons appear to "disappear" as they pass in front of the planet? Which moons "disappear" as they pass behind the planet?

2. Consult a current astronomy magazine to locate Saturn's position and then go out at night and try to observe the planet. Good binoculars or a small telescope should also permit you to observe Saturn's rings and its large moon, Titan. Make a sketch of your observations. Try to include Cassini's division in your sketch. (You may also use the coordinates in Lesson 21 in this guide and the SC1 Star Chart.)

3. With the aid of a small telescope with an aperture of at least 7 centimeters (3 inches), or very good binoculars, observe the large moon, Titan, over a period of a month. Initially, it may be difficult to distinguish Titan from the background stars that appear near Saturn. However, by sketching Saturn on successive nights and locating the nearby stars as seen in the telescope, and then comparing the sketches, you will observe two objects—Saturn and Titan—appearing to move together relative to the stationary background stars. Trace their paths relative to the background stars and relative to each other. From your sketch, you should be able to estimate the period of Titan's revolution around Saturn. How does this period compare with the revolutions of the Galilean satellites and why is Titan's period different?

Lesson Review
Lesson 23
Jupiter and Saturn

Please Note: Use this matrix to guide your study and achieve the learning objectives of this lesson. It will also help you to view the video, which defines and demonstrates important concepts and skills as they relate to everyday life.

Learning Objective	Textbook	Telecourse Student Guide	Background Notes
1. Describe the size, shape, rotation, and orbital characteristics of Jupiter and Saturn and describe their appearance as seen through a telescope.	pp. 381–388, 391–392, 394–395	Matching Activities: 2, 6; Self-Test Questions: 1, 2, 3, 4.	
2. Recount the discoveries made by the Voyager and Galileo spacecraft at Jupiter and the Voyager spacecraft at Saturn.	pp. 381–397	Matching Activities: 1, 2; Self-Test Questions: 5, 6, 20.	
3. Identify the Galilean satellites in orbit around Jupiter, provide details of their physical characteristics, and explain the origin of the identified features.	pp. 388–391	Matching Activities: 3, 4, 5; Completion Activities: 1; Self-Test Questions: 7, 8.	
4. Compare and contrast the ring systems of Jupiter and Saturn and explain the origin of these ring systems.	pp. 386–388, 392, 394–395	Matching Activities: 8; Self-Test Questions: 9, 10, 11.	
5. Describe the current models of the atmosphere and interior of Jupiter and Saturn and explain the similarities and differences.	pp. 382–386, 391–393	Self-Test Questions: 12, 13.	Background Notes 23: The Galileo Probe into the Atmosphere of Jupiter
6. Identify the major moons in orbit around Saturn, describe their physical characteristics, and explain the origin of these features.	pp. 392–397	Self-Test Questions: 14, 15.	

Learning Objective	Textbook	Telecourse Student Guide	Background Notes
7. Describe in detail the characteristics of the extensive ring system around Saturn and explain its structure.	pp. 392–395	Matching Activities: 7; Completion Activities: 2 Self-Test Questions: 16, 17.	
8. Compare and contrast the characteristics and physical appearance of the large storms (spots) on Jupiter and Saturn.	pp. 385–386, 391	Matching Activities: 18, 19.	

L E S S O N
24

Uranus, Neptune, and Pluto

Assignments

For the most effective study of this lesson, we suggest that you complete the assignments in the sequence listed below.

Before Viewing the Video Program

- Read the Overview and Learning Objectives for this lesson. Use the Learning Objectives to guide your reading, viewing, and thinking.

- Read Chapter 18, "Worlds of the Outer Solar System," sections 18-3 through 18-5, pages 397–408, in the *Horizons* textbook.

- Read the Viewing Notes in this lesson.

View the "Uranus, Neptune, and Pluto" Video Program

After Viewing the Video Program

- Briefly note your answers to questions listed at the end of the Viewing Notes.

- Review all reading assignments for this lesson, especially the Chapter 18 summary on pages 408–409 in *Horizons* and the Viewing Notes in this lesson.

- Respond to the challenges presented by each Critical Inquiry in the reading assignment and write brief answers to review questions 9 and 11–15 at the end of Chapter 18 in *Horizons* to be certain you understand the text material.

- Complete the Review Activities in this guide to reinforce your understanding of important terms and concepts. Check your answers with the Answer Key and review when necessary.

- Take the Self-Test in this guide to measure your achievement of the Objectives. Check your answers with the Answer Key and review when necessary.

- Complete the Using What You've Learned activities and any other activities and projects assigned by your instructor.

Overview

This lesson on Uranus, Neptune, and Pluto takes us to the edge of our solar system and completes our study of the planets. Much of what astronomers know about two of these most distant planets—Uranus and Neptune—was learned quite recently from data and images sent back by Voyager spacecraft.

Of course, the information provided by the Voyager missions had been valuable in studying the other planets, but it was exceptionally valuable for learning more about Uranus and Neptune, because the great distances to those planets had impeded their study with Earth-based telescopes.

Astronomers were surprised and challenged by much of the information sent back by Voyager: volcanoes on Io, ringlets around Saturn, shepherding moons. Information about Uranus and Neptune was even more surprising and challenging. The spacecraft saw dark rings around Uranus and, passing Neptune, Voyager sent back striking images of clouds and a dark spot and rings composed of dark material.

Even the moons around Uranus and Neptune surprised astronomers. One moon around Uranus has a number of significantly different geologic features, while others have regions on their surfaces that show very little cratering. The moon Triton, orbiting Neptune in nearly the coldest region of the solar system, is volcanically active. The question for astronomers, of course, was: Where does the heat come from? Even Pluto, the smallest planet in the solar system, and its moon Charon, raised questions for astronomers about the history and formation of the solar system. Astronomers are debating whether Pluto, Charon, and Triton are a new class of objects orbiting in the farthest reaches of the solar system, and if collisions with these kinds of objects caused the peculiar behavior of Uranus.

All the information gathered about the planets and associated objects and studied by astronomers is important in developing and refining theories on the formation, history, and evolution of the solar system. This knowledge is important, not only because knowledge is important for its own sake but also because it helps to put planet Earth in perspective. Earth is a planet orbiting around a relatively common star. It exists in an environment with other planets and minor objects. By studying the entire solar system, astronomers can develop an understanding of how this environment affects the planet and the life on its surface.

Learning Objectives

When you have completed all assignments for this lesson, you should be able to:

1. Describe the orbital, rotational, and physical characteristics of Uranus, Neptune, and Pluto as observed from Earth and compare these characteristics with Mercury, Venus, Earth, Mars, Jupiter, and Saturn. *HORIZONS* TEXTBOOK PAGES 397–399, 401–404, AND 406–408; VIDEO PROGRAM.

2. Describe the events leading to the discoveries of Uranus, Neptune, and Pluto. VIDEO PROGRAM.

3. Compare and contrast the characteristics of the atmospheres of Uranus and Neptune as observed by the Voyager spacecraft. *HORIZONS* TEXTBOOK PAGES 398–399, 401–405, AND Figure 18-16; VIDEO PROGRAM.

4. Describe the events leading to the discoveries of the ring systems around Uranus and Neptune and explain how astronomers deduced that rings existed around Uranus and Neptune. *HORIZONS* TEXTBOOK PAGES 400, 402–403, AND 405; VIDEO PROGRAM.

5. Describe the features of the ring systems of Uranus and Neptune and explain the origin of these features. *HORIZONS* TEXTBOOK PAGES 400, 402–403 AND 405.

6. Describe the physical characteristics of the moons of Uranus and Neptune and explain the origins of these features. *HORIZONS* TEXTBOOK PAGES 400–401 AND 405–406; VIDEO PROGRAM.

7. Compare and contrast the orbital characteristics of the satellite systems of Uranus, Neptune, and Pluto. *HORIZONS* TEXTBOOK PAGES 398–400 AND 405.

8. Describe the known physical characteristics of Pluto and its moon Charon. *HORIZONS* TEXTBOOK PAGES 406–408; VIDEO PROGRAM.

Viewing Notes

As with the other video programs on the planets, "Uranus, Neptune, and Pluto" provides descriptions of each planet, accompanied by NASA and JPL photographs.

The segment on Uranus begins by relating the story of how William Herschel accidentally discovered the planet in 1781. Astronomers go on to describe the planet's composition, the tip in its orbital axis, its rotation, magnetic field, and atmosphere. The video program also supplies information about Uranus' dark rings and the planet's moons, including Miranda and theories about that moon's unusual history.

In discussing Neptune in the next segment, astronomers explain that planet's similarities to Uranus and note important differences. The atmosphere of Uranus is more active, and the planet has white clouds and a great dark spot. The video program also supplies information on Neptune's rings and Triton, one of its two major moons. Planetary scientists describe some of icy Triton's unusual features, including its retrograde orbit, rocky interior, nitrogen and methane atmosphere, relative absence of craters, and dark "smudges," which may be caused by active volcanoes or geysers.

The last part of the video program focuses on Pluto, the most distant planet in the solar system, and its moon, Charon. Pluto is also the most recently discovered planet, and the story of its discovery by Clyde Tombaugh in 1930 is related. Because of its great distance, little is known about Pluto, but astronomers explain how they have been able to determine the planet's mass and how spectroscopy has provided information on the planet's composition. Planetary scientists describe current theories about the origin of this unusual planet as well as the origin of similar icy bodies in the distant areas of the solar system. Finally, astronomers briefly discuss the possibility of discovering other planets beyond Pluto.

As you watch the video program, consider the following questions.

1. Why is the disk of Uranus so featureless?

2. How were the rings of Uranus discovered?

3. What constituent of the atmosphere of Neptune makes it appear blue?

4. What feature on Neptune is comparable to the Great Red Spot on Jupiter?

5. How was Pluto discovered?

6. How was its moon Charon discovered?

Review Activities

Matching

Match the terms listed below with the definitions that follow. Check your answers with the Answer Key and review any terms you missed.

_____ 1. methane

_____ 2. occultation

_____ 3. ovoids

_____ 4. ring arcs

_____ 5. retrograde orbit

_____ 6. Charon

a. backward motion of satellite as it orbits a planet

b. oval features found on Miranda

c. sections of complete rings, unique to planet Neptune

d. passage of one celestial body in front of another

e. Pluto's moon

f. gas that makes the atmospheres of Uranus and Neptune appear blue

Matching—Planet Review

Now that you have completed your study of the planets of our solar system, the following activity will help you review what you have learned about these worlds. Match each description or feature in the numbered list below with the planet or planets to which it applies from the list that follows. Check your answers with the Answer Key and review any planetary features you missed.

_____ 1. largest number of known satellites

_____ 2. famous red spot

_____ 3. surface with the highest temperature

_____ 4. frozen water and evidence of former rivers

_____ 5. smallest mass

_____ 6. largest mass

_____ 7. farthest from Earth

_____ 8. polar caps

_____ 9. strongest magnetic field

_____ 10. third from the sun

_____ 11. rings

_____ 12. most like Earth in size

_____ 13. like Earth in surface conditions

_____ 14. closest to the sun

_____ 15. most recently discovered

_____ 16. discovered by mathematical prediction

_____ 17. associated with Cassini's division

_____ 18. largest amount of carbon dioxide in its atmosphere

_____ 19. lowest mean density

_____ 20. orbited by Charon

a. Jupiter	e. Pluto	i. Neptune
b. Saturn	f. Earth	j. none
c. Venus	g. Mercury	
d. Mars	h. Uranus	

Completion

Fill each blank with the most appropriate term from the list for that paragraph. A term may be use once, more than once, or not at all. Check your answers with the Answer Key and review when necessary. If a question requires two or more answers in succession, they may be in any order, unless the question indicates otherwise.

1. The last two giant planets of the outer solar system have significant similarities and differences when compared to Jupiter and Saturn. Both Uranus and Neptune are composed mostly of _____ and _____ just as are Jupiter and Saturn. Uranus and Neptune, however, do not have sufficient densities or pressures to produce an interior of liquid _____ hydrogen. The interiors of these worlds are simply _____ hydrogen. The cloud patterns of Uranus and Neptune are also noticeably different from those of Jupiter and Saturn. Through a telescope, both Uranus and Neptune appear to be _____ and they do not have the colorful _____ and _____ so visible in the clouds of Jupiter. In fact, Uranus is practically _____. The apparent color of Uranus and Neptune is due to the presence of _____ in the atmosphere. This gas absorbs _____ and _____ light, thus producing the _____ coloring.

   When astronomers first saw the images of Neptune taken by Voyager, they were surprised. Because of Neptune's great distance and very low _____, they had expected the planet to appear _____ and _____. However, Neptune showed _____ detail than Uranus and even had a feature similar to the Great Red Spot on Jupiter. The color of this feature is _____ and it is referred to as the _____. Astronomers concluded that Neptune must be producing some internal _____, causing _____. Combined with the _____ rotation of the planet, _____ winds and _____ circulations are generated.

bands	high-speed
belts	hydrogen
bland	less
blue	liquid
blue-green	metallic
cold	methane
convection	more
cyclonic	rapid
featureless	red
Great Dark Spot	slow
heat	temperatures
helium	yellow

2. The rings of Uranus and Neptune are very much different from the rings of Saturn. The rings of Uranus were detected when the planet _____ a star. Dips in the _____ from the star appeared _____ of the planet and a _____ pattern of dips appeared _____ the planet. The rings of Uranus are very _____. The structure of these rings depends on the presence of tiny _____, which control the position of the ring _____. These objects are called _____. Unlike the rings of Saturn, the rings of Uranus appear _____. It has been hypothesized that this color of the rings is due to the presence of _____. These chemicals were produced when the frozen _____ in the ring particles was converted into this _____ matter by the radiation associated with the _____ of Uranus.

ahead	narrow
behind	occulted
black	organic polymers
broad	particles
dark	radiation belts
light	shepherding satellites
methane	symmetrical
moons	

3. Because Pluto's orbit is highly _____, that planet is sometimes_____ to the sun than is Neptune. The two planets will never collide, because the orbit of Pluto is _____ to the orbit of Neptune, and the two objects will never be at the same place in the solar system at the same time. Because of its great distance, very little else was known about Pluto until the discovery of a moon in orbit about the planet.

The discovery of this moon, called _____, enabled astronomers to calculate the _____ of Pluto. By analyzing these two objects as they would an _____ star system, astronomers were able to determine the _____ of both objects. The size of this moon is about _____ that of Pluto. Calculations of the density of Pluto and the moon revealed they are composed mostly of a mixture of _____ and _____.

There is a highly speculative explanation for the origin of Pluto. Because of the density and size calculations that have been made, Pluto and its moon seem very similar to the moon _____. This conclusion leads scientists to believe that there are other _____ bodies beyond the orbit of Pluto. These objects have been tentatively called _____. Some astronomers think that a near-collision of Uranus with these objects caused a severe tilt to the planet's axis of _____ and the moon Triton's orbit around Neptune to be _____.

Callisto	mass
Charon	moon
closer	one-half
diameters	one-third
eclipsing binary	retrograde
elliptical	rocks
ice	rotation
icy	Triton
icy planetesimals	Uranus
inclined	

Self-Test

Select the one best answer for each question. Check your answers with the Answer Key and review when necessary.

1. Which of the following planets has its axis of rotation tipped the most?
   a. Uranus
   b. Neptune
   c. Pluto
   d. Charon

2. Which of the following planets has the most elliptical orbit?
   a. Uranus
   b. Triton
   c. Neptune
   d. Pluto

3. The discovery of Neptune was made as the result of
   a. a study of planetary nebula.
   b. a search for comets.
   c. predictions following mathematical calculations.
   d. a search for near-Earth asteroids.

4. Percival Lowell believed that the irregularities in the motion of the planet Uranus were caused by a planet beyond Neptune. His efforts and the efforts of others that followed led to the eventual discovery of
   a. Triton.
   b. Nereid.
   c. Pluto.
   d. Charon.

5. The very bland appearance of the atmosphere of Uranus is probably caused by

   a. the absence of hydrogen and helium.

   b. extremely cold temperatures and limited heat flow from its interior.

   c. the absence of methane in the atmosphere.

   d. the presence of methane in the atmosphere.

6. Neptune's blue appearance is the result of the presence of

   a. hydrogen.

   b. helium.

   c. ammonia.

   d. methane.

7. The rings of Uranus were discovered during

   a. the Voyager spacecraft flyby.

   b. the occultation of a star by Uranus.

   c. an eclipse of Uranus by the moon Miranda.

   d. an occultation of Uranus by Saturn.

8. The rings of Neptune were discovered as the result of an occultation of a star by Neptune. Astronomers concluded that the rings were not complete because

   a. only one side of the planet yielded the expected dimming of the starlight.

   b. only half the rings were visible from Earth.

   c. theoretical calculations predicted these results.

   d. only half the rings were observed by the Voyager spacecraft.

9. The ring systems of Uranus and Neptune consist of very narrow rings. This narrowness is probably the result of

   a. random collisions that have driven particles out of the ring system.

   b. magnetic constrictions resulting from the interaction of the particles with the magnetic fields of the planets.

   c. shepherding satellites controlling the positions of the ring particles.

   d. solar radiation pressure.

10. The rings of Uranus are extremely dark. This appearance is probably caused by the presence of

    a. volcanic ash.

    b. radiation from Uranus's radiation belts breaks down the methane ices and releases carbon that darkens the ring particles.

    c. frozen methane.

    d. inorganic ammonia compounds.

11. The short arcs in the narrow outer ring of Neptune can be explained by a mathematical model that shows

    a. a gravitational resonance with the moon Triton.

    b. shepherding satellites.

    c. a small (as yet unobserved) moon moving in and out of the ring, sweeping portions of the ring clear of particles.

    d. waves in the ring produced by the inner moon Galatea, confining the particles into small arcs.

12. The oval patterns of grooves (ovoids) on the surface of Uranus's moon Miranda is explained by

    a. gravitational influences produced by Uranus.

    b. tidal influences produced by the other nearby moons Umbriel and Ariel.

    c. collisions between Miranda and the ring particles.

    d. internal heat that produced convection in the icy mantle.

13. Dark smudges on the surface of Neptune's moon Triton appear to be the result of

    a. organic polymers resulting from ammonia interacting with Neptune's magnetic field.

    b. impact debris from asteroid collisions.

    c. nitrogen volcanism.

    d. disintegration of ring particles falling on the planet.

14. Neptune's moon Triton
    a. is in a retrograde orbit.

    b. is in a perfectly circular orbit.

    c. orbits in the same direction as the other moon, Nereid.

    d. has a polar orbit.

15. Pluto and Charon—and Neptune's largest moon, Triton—are small worlds that may be the remains of many such ice bodies, called _____, that once populated the outer solar system.
    a. Plutons

    b. ovoids

    c. planetesimals

    d. protoplanets

16. Pluto's moon Charon is approximately
    a. half the size of Pluto.

    b. twice the size of Pluto.

    c. one-fourth the size of Pluto.

    d. equal to the size of Pluto.

17. The mass of Pluto was determined
    a. by direct observation from Voyager.

    b. from the size of the orbit and period of its moon, Charon.

    c. by direct observation from Earth-based telescopes.

    d. by direct observation from the Hubble Space Telescope.

Using What You've Learned

1. Describe the seasons as they would occur on Uranus, and compare them to those on Earth.

2. Outline the events that led to the discoveries of Uranus, Neptune, and Pluto. What role did serendipity play?

3. Explain why Pluto, at times, is not the most distant planet in the solar system.

Lesson Review
Lesson 24
Uranus, Neptune, and Pluto

Please Note: Use this matrix to guide your study and achieve the learning objectives of this lesson. It will also help you to view the video, which defines and demonstrates important concepts and skills as they relate to everyday life.

Learning Objective	Textbook	Telecourse Student Guide	Background Notes
1. Describe the orbital, rotational, and physical characteristics of Uranus, Neptune, and Pluto as observed from Earth and compare these characteristics with Mercury, Venus, Earth, Mars, Jupiter, and Saturn.	pp. 397–399, 401–404, 406–408	Matching Activities: 1, 3; Matching—Planet Review: 1, 2, 3, 4, 5, 6, 7, 8, 9, 10, 11, 12, 13, 14, 15, 16, 17, 18, 19, 20; Completion Activities: 1; Self-Test Questions: 1, 2.	
2. Describe the events leading to the discoveries of Uranus, Neptune, and Pluto.	pp. 381–397	Self-Test Questions: 3, 4.	
3. Compare and contrast the characteristics of the atmospheres of Uranus and Neptune as observed by the Voyager spacecraft.	pp. 398–399, 401–405; Figure 18-16	Completion Activities: 1; Self-Test Questions: 5, 6.	
4. Describe the events leading to the discoveries of the ring systems around Uranus and Neptune and explain how astronomers deduced that rings existed around Uranus and Neptune.	pp. 400, 402–403, 405	Matching Activities: 2; Self-Test Questions: 7, 8.	
5. Describe the features of the ring systems of Uranus and Neptune and explain the origin of these features.	pp. 400, 402–403, 405	Matching Activities: 4; Completion Activities: 2; Self-Test Questions: 9, 10, 11.	
6. Describe the physical characteristics of the moons of Uranus and Neptune and explain the origins of these features.	pp. 400–401, 405–406	Self-Test Questions: 12, 13.	

Learning Objective	Textbook	Telecourse Student Guide	Background Notes
7. Compare and contrast the orbital characteristics of the satellite systems of Uranus, Neptune, and Pluto.	pp. 398–400, 405	Matching Activities: 5, 6; Completion Activities: 3; Self-Test Questions: 14, 15.	
8. Describe the known physical characteristics of Pluto and its moon Charon.	pp. 406–408	Matching Activities: 6; Completion Activities: 3; Self-Test Questions: 16, 17.	

Meteorites, Asteroids, and Comets

Assignments

For the most effective study of this lesson, we suggest that you complete the assignments in the sequence listed below.

Before Viewing the Video Program

- Read the Overview and Learning Objectives for this lesson. Use the Learning Objectives to guide your reading, viewing, and thinking.

- Read Chapter 19, "Meteorites, Asteroids, and Comets," pages 411–429, in the *Horizons* textbook.

- Read Background Notes 25A, "Comet Halley—History"; 25B, "Comet Shoemaker-Levy 9"; and 25C, "Interplanetary Debris Impacts," following Lesson 26 in this guide.

- Read the Viewing Notes in this lesson.

View the "Meteorites, Asteroids, and Comets" Video Program

After Viewing the Video Program

- Briefly note your answers to questions listed at the end of the Viewing Notes.

- Review all reading assignments for this lesson, especially the Chapter 19 summary on pages 428 in *Horizons*, Background Notes 25A–25C for Unit IV, and the Viewing Notes in this lesson.

- Respond to the challenges presented by each Critical Inquiry in the reading assignment and write brief answers to the review questions at the end of Chapter 19 in *Horizons* to be certain you understand the text material.

- Complete the Review Activities in this guide to reinforce your understanding of important terms and concepts. Check your answers with the Answer Key and review when necessary.

- Take the Self-Test in this guide to measure your achievement of the Objectives. Check your answers with the Answer Key and review when necessary.

- Complete any other activities and projects assigned by your instructor.

Overview

After learning about the planets of our solar system, you may have a tendency to think of the smaller objects in the solar system as "leftovers." These objects range in size from that of a very small moon down to dust particles. Some of these objects—comets and meteors—are responsible for startling visual effects. Others, such as asteroids, are invisible to the unaided eye and could be overlooked even if photographed through a telescope. Why does all this debris attract so much attention? There aren't any definite answers yet concerning the solar system's origins, but these objects and particles do contain intriguing clues about how the solar system was formed.

This lesson is concerned with the organization, classification, and analysis of the smallest pieces of the solar system. The "rocks" from outer space are correctly called meteorites and can be classified according to their appearance and composition. Astronomers now know that meteorites come from larger parent bodies: asteroids. Asteroids, sometimes called "minor planets," are the remaining planetesimals from which the planets and moons were made. Detailed study of meteorites helps scientists to learn about the history and early processes through which the asteroids have passed. The link between meteorites and asteroids allows scientists to look back into the earliest times of the solar system.

Unfortunately, most of the meteorites that arrive on Earth have been altered through heating or melting. Although these processes occurred in the earliest times of the solar system, meteorites, except for one type, do not reveal much about the original matter of the solar system.

In contrast to meteorites and asteroids, comets are the most pristine objects in the solar system. Made of ice and rock, any melting or heating would have caused comets to be obliterated within the 4.5 billion years of the solar system's existence. Since comets still exist and occasionally visit the vicinity of Earth, they must exist in a region of the solar system that is free from the high temperatures that would change their structures. In this outermost region of the solar system, scientists may one day find answers to the questions about the origin of the planets, their moons, and possibly life itself.

In the past, these "hairy stars" created much superstition and fear, and it was not until the nineteenth century when astronomers began to learn about comets and discredit some of the pseudoscientific myths that were associated with a comet's apparition.

The study of comets was Earth-based until the return of Comet Halley in 1986. At that time, spacecraft were launched to rendezvous with Halley. These spacecraft took photographs of the comet's nucleus, measured its characteristics, and sampled the contents of its tail. The encounter with Comet Halley provided new information about the shape, chemical composition, and the possibility that the crust of the comet is composed of carbon. Comets, once believed to be omens of death and destruction, could very well be the source of Earth's first oceans and material that eventually allowed life to form and evolve on this planet. The smallest pieces within the solar system may tell us the biggest story.

Learning Objectives

When you have completed all assignments for this lesson, you should be able to:

1. Identify the three broad categories of meteorites and the three types of stony meteorites. *HORIZONS* TEXTBOOK PAGES 413–414; VIDEO PROGRAM.

2. Describe the possible origins of the different types of meteorites. *HORIZONS* TEXTBOOK PAGES 414–416; VIDEO PROGRAM.

3. Describe the characteristics of meteors and meteor showers, and explain how they relate to cometary orbits. *HORIZONS* TEXTBOOK PAGES 414–415.

4. Describe the physical properties and orbital characteristics of asteroids. *HORIZONS* TEXTBOOK PAGES 416–418; VIDEO PROGRAM.

5. Explain theories about the origin of asteroids. *HORIZONS* TEXTBOOK PAGE 419; VIDEO PROGRAM.

6. Identify and describe the major parts of a comet and explain the effect of solar wind on the tail of a comet. *HORIZONS* TEXTBOOK PAGES 420–424; VIDEO PROGRAM.

7. Explain the significance of the Oort cloud and Kuiper belt theories to understanding the origin of comets. *HORIZONS* TEXTBOOK PAGES 421, 424; VIDEO PROGRAM.

8. Describe the significant historical observations of Comet Halley, including the 1910 apparition, and the important discoveries made by spacecraft during the 1986 apparition. *HORIZONS* TEXTBOOK PAGES 420–421; BACKGROUND NOTES 25A; VIDEO PROGRAM.

9. Explain the possible effects of an asteroid/comet collision with Earth and describe the impact of Comet Shoemaker-Levy 9 on Jupiter. *HORIZONS* TEXTBOOK PAGES 424–427; BACKGROUND NOTES 25B AND 25C; VIDEO PROGRAM.

Viewing Notes

This video program concludes our exploration of the solar system with an examination of solar system "debris": meteorites, asteroids, and comets.

The segment on meteorites features astronomers explaining the composition of meteorites and various theories about their origin. The program also includes examples of the major types of meteorites.

Asteroids, thought to be the source of meteorites, are covered in the next segment. The video program provides information on the three classifications of asteroids—S types, C types, and M types—and the relationship of each classification to the different types of meteorites found on Earth's surface. Astronomers describe how the first asteroid, Ceres, was discovered in 1801. It was not until spacecraft carried telescopes into space that astronomers were able to learn about the surface features and shape of these small objects. In the program, you will see a photograph of the asteroid Gaspra, taken by the Galileo spacecraft in 1990.

Astronomers describe how most asteroids orbit in a belt between Mars and Jupiter and offer theories about the origin of asteroids. They also provide information on asteroids outside the main asteroid belt. Trojan asteroids precede and follow Jupiter in its orbit, and Apollo asteroids have elongated orbits that cross Earth's orbit. Some Apollo asteroids are in orbits that occasionally bring them close to Earth. Ultimately, these asteroids crash into a planet. The video program looks at a possible asteroid or comet impact crater near the Yucatán, in Mexico, and considers the theory that the aftereffects of the impact led to the extinction of the dinosaurs.

Comets, some of which produce spectacular celestial events as they arc through the sky, are the subject of the final segment. The segment includes information on Comet Halley, probably the most famous of all comets seen from Earth, and examines the lore surrounding some of the comet's past apparitions. In the program, you'll see photographs of Comet Halley taken by the spacecraft Giotto during the comet's most recent appearance in 1985–1986. These photographs supplied astronomers with valuable information about the comet's nucleus, and astronomers describe what is now known about a comet's head and tail and composition. They offer two principal theories—the Oort cloud theory and the Kuiper belt theory— about the origin of comets. This segment also includes footage of a most spectacular cometary event: the impact of Comet Shoemaker-Levy 9 on Jupiter in 1994. Staff from the Jet Propulsion Laboratory describe the collision of the comet's 20 fragments and offer some preliminary findings.

As you watch the video program, consider the following questions:

1. How do meteors, meteoroids, and meteorites differ?

2. What are the different types of meteorites and what kinds of information do these objects reveal?

3. What is an asteroid and where is the asteroid belt?

4. What are the different classes of asteroids and how are these classes related to the different types of meteorites?

5. What is a comet and what characteristics of a comet are visible through a telescope?

6. What are the parts and composition of a comet?

7. Where do comets originate?

8. What are the possible effects of a cometary impact on Earth?

9. Why is Comet Halley so important and what was learned during its last visit to the inner solar system?

Review Activities

Match the terms below with the definitions that follow. Check your answers with the Answer Key and review any terms you missed.

Matching

_____ 1. achondrite

_____ 2. carbonaceous chondrite

_____ 3. chondrite

_____ 4. chondrule

_____ 5. coma

_____ 6. meteor shower

_____ 7. Oort cloud

_____ 8. Widmanstätten pattern

_____ 9. Kuiper belt

_____ 10. radiant point

_____ 11. meteor

_____ 12. meteorite

_____ 13. meteoroid

a. bands in iron meteorites created by large crystals of nickel-iron alloys

b. a stony meteorite with chondrules

c. round, glassy body found in some stony meteorites

d. stony meteorite with both chondrules and volatiles

e. stony meteorite with composition similar to that of Earth's lavas

f. event that occurs when Earth passes near the orbit of a comet, probably caused by dust and debris released by a comet

g. glowing head of a long-period comets

h. hypothetical source of long-period comets

i. small bit of matter heated by friction to incandescent (glowing) vapor as it falls into Earth's atmosphere

j. region beyond the orbit of Neptune that is a reservoir for short-period comets

k. small object or particle in space before it enters Earth's atmosphere

l. small region in the sky from which meteor showers appear to originate

m. small object or particle that survives passage through Earth's atmosphere and reaches the ground

Completion

Fill each blank with the most appropriate term from the list for that paragraph. A term may be used once, more than once, or not at all. Check your answers with the Answer Key and review when necessary. If a question requires two or more answers in succession, they may be in any order, unless the question indicates otherwise.

1. At certain times of the year, a patient observer can see flashes of light in the night sky. These flashes of light have been incorrectly called _____. These flashes are produced by small particles or objects and are correctly called _____. The particles themselves are called _____. Periodically, one of the particles is large enough to survive the passage through Earth's atmosphere and reach Earth's surface. Such an object is called a _____.

    Upon careful analysis, these objects can be classified into three broad categories. The first group is made of metals and is called _____. In fact, these objects are made of _____ and _____. When these objects are cut in half, polished, and etched with _____ acid, regular bands appear. These bands are called _____ patterns. The bands were produced when the object cooled very _____ and can be used in trying to understand the _____ of the object.

    The second group of objects looks very much like Earth rocks. These objects are called _____. This second group can be subdivided into three additional groups. The objects with small,

rounded bits of glassy rock imbedded within are known as
_____. The bits of glass are called _____. The
second subgroup also contain these bits of glass, along with carbon
and water. These _____ can exist only if the object has not
been heated or melted in any way. This second group, known as
_____, contains the least altered remains of the solar nebula.
Finally, the objects in the third group of stony meteorites contain no
glass or _____. They appear to have been subjected to intense
_____ and look very similar to Earth _____. This
last group is called the _____.

The third type of meteorites is called _____. These are a
mixture of _____ and _____. They appeared to have
been formed when a mixture of _____ iron and
_____ solidified from a _____ state.

achondrites	molten
carbonaceous chondrites	nickel
chondrites	nitric
chondrules	rapidly
cooling	rock
heating	shooting stars
history	slowly
iron	stone
iron meteorites	stony-iron
lava	stony meteorites
meteorites	volatiles
meteoroids	Widmanstätten
meteors	

2. In 1772, a rule predicting the distances to the known planets was
announced. Known as the Titus-Bode rule, it also predicted the
existence of a planet between the orbits of _____ and
_____ where one was not known to exist. Although no planet

was ever discovered in that region, astronomers discovered many smaller objects. These objects are known as _____. Studies of the color and spectra of these objects show that they can also be divided into three groups.

The S-type objects are very _____ and their color appears _____. They seem to be silicates mixed with _____ and may be associated with meteorites from the _____ group. The C type objects are very dark and seem to be _____ and associated with the meteorites of the _____ class. A further connection between this group and the meteorites is that these objects are located in the _____ belt, indicating that they have been exposed to extremely _____ temperatures and have experienced very little _____ or melting. Finally, the M-type objects are very bright but without the _____ color of the _____ objects. Available evidence indicates that they are made of metals, probably _____ and _____. These objects are similar to _____.

asteroids	Jupiter
blue	low
bright	M-type
C-type	Mars
carbonaceous	metals
carbonaceous chondrites	Neptune
chondrites	nickel
cooler	outer
heating	Pluto
high	red
inner	S-type
iron	ultraviolet
iron meteorites	

3. To early humans, one of the most frightening events occurring in the sky was the appearance of a comet. These objects looked like a fuzzy ball of light with a very long _____. It wasn't until the nineteenth century that science began to understand the nature of these objects. A comet can be described as a mountain of ice and rock. In fact, many used to call the central lump, or _____, a dirty _____. However, study of Halley's nucleus from photographs taken in 1986 reveals a density _____ than that of ice. This result indicates the nucleus is a _____ mixture of ices and a _____ dusty crust. Vaporization of the ices in this central lump forms the _____. This large cloud of _____ and _____ is what is visible when a comet is viewed through a telescope. Emerging from this vast cloud is the tail of the comet, which may be _____ of kilometers long.

Comets have two kinds of tails. The first is straight and wispy and is a _____ tail. It is made up of _____ gases and it is blown _____ by moving gas and the embedded magnetic field of the solar wind. The second kind of tail is the _____ tail, which is unaffected by magnetic fields. This tail reflects the spectrum of sunlight and is therefore made up of _____ bits of _____ from the central lump. Although the material in this tail does not respond to magnetic effect, it is pushed _____ by the pressure of _____. It is interesting to note that the tail of a comet is always directed _____ from the sun.

away	liquid
billions	millions
coma	nucleus
dark	outward
dust	snowball
fluffy	solid
gas	sunlight
ionized	tail
less	

Self-Test

Select the one best answer for each question. Check your answers with the Answer Key and review when necessary.

1. Which of the following are classifications of meteorites?
    a. basalt and igneous
    b. irons and magnetite
    c. iron and stony
    d. shower and crystal

2. Widmanstätten patterns are found in
    a. achondrites.
    b. iron meteorites.
    c. chondrules.
    d. stony meteorites.

3. Because of the delicate nature of the carbonaceous chondrites, it is believed that these meteorites came from planetesimals that
    a. orbited the sun near the planet Mercury.
    b. collided with Earth billions of years ago.
    c. orbited far from the sun in regions of the solar system that are very cold and are probably the least altered remains of the solar nebula.
    d. collided with Mars, causing material to be ejected from the planet and eventually land on Earth.

4. The iron meteorites probably formed from
    a. large planetesimals that melted and differentiated to form an iron core.
    b. condensing iron atoms during the cooling period of the early solar nebula.
    c. collisions with Mars-sized planetesimals.
    d. cometary impacts with asteroids.

5. A meteor is

    a. a particle from outer space.

    b. a particle from outer space that lands on Earth.

    c. produced by tiny bits of debris from comets.

    d. a small asteroid.

6. A meteor shower occurs when

    a. an asteroid collides with Earth.

    b. Earth passes through the dust and bits of material left in the orbit of a passing comet.

    c. meteorites are found in a short period of time.

    d. Earth reaches its nearest position to the sun (perihelion).

7. Most of the asteroids revolve about the sun between the orbits of

    a. Jupiter and Saturn.

    b. Mars and Jupiter.

    c. Earth and Mars.

    d. Saturn and Uranus.

8. M-type asteroids appear to

    a. orbit in a fixed position between Jupiter and Mars.

    b. be made of stone and silicates and are red in color.

    c. consist of carbon and look very dark.

    d. be made of iron-nickel alloy and are very bright.

9. The formation of asteroids is the result of

    a. a collision between two planets.

    b. planetary material that never formed into large planets.

    c. an exploded planet.

    d. the escape of Jovian-type satellites.

10. Many collisions caused planetesimals to fragment into smaller pieces. These pieces probably make up the material known as

    a. meteor showers.

    b. meteorites.

    c. asteroids.

    d. planetary moons.

11. The solar wind directs the tail of a comet

    a. outward, away from the sun at all times.

    b. away from the sun when the comet enters the solar system and toward the sun when it leaves the solar system.

    c. toward the sun when the comet enters the solar system and away from the sun when it leaves.

    d. toward the sun at all times.

12. The nucleus of the comet contains mostly the ices of water, carbon dioxide, and ammonia. The coma is produced when the

    a. nucleus explodes.

    b. comet reaches its most distant point from the sun (aphelion) and the ices vaporize.

    c. tail of the comet compresses back toward the nucleus.

    d. sun causes sufficient heating and the ices of the nucleus vaporize to produce a vast cloud of gas and dust.

13. The source of the long-period comets is a region of space at least 10,000 astronomical units from the sun. This region of space is called

    a. interstellar space.

    b. the interstellar medium.

    c. the Oort cloud.

    d. the Kuiper belt.

14. Studies show that the short-period comets cannot originate at a distance thousands of astronomical units from the sun. A closer source for short-period comets is the

   a. Kuiper belt.

   b. Oort cloud.

   c. Lagrangian points.

   d. asteroid belt.

15. Comet Halley was first observed

   a. in 1910.

   b. more than 2,000 years ago.

   c. during the Norman invasion of A.D. 1066.

   d. by Edmund Halley and Isaac Newton.

16. The spacecraft that intercepted Comet Halley in 1985 and 1986 discovered that the nucleus of Comet Halley is

   a. perfectly round and smooth.

   b. bright and irregularly shaped.

   c. very dark and irregularly shaped.

   d. very dark and perfectly smooth, with no obvious relief.

17. Jets issue from the nucleus of Comet Halley only

   a. from the side of the comet pointing away from the sun.

   b. when the comet is at aphelion.

   c. during the period of time when the tail is visible.

   d. from the sunlit side of the comet.

18. Evidence that Earth has collided with an asteroid-sized object is suggested by a layer of the rare element

   a. ytterbium.

   b. plutonium.

   c. iridium.

   d. planetarium.

19. The "smoking gun" crater that fits the description of the asteroid impact that may have caused the extinction of the dinosaurs was found

   a. off the coast of Hawaii.

   b. off the coast of the Yucatán peninsula.

   c. outside of Flagstaff, Arizona.

   d. in Crater Lake, Oregon.

Lesson Review

Lesson 25

Meteorites, Asteroids, and Comets

Please Note: Use this matrix to guide your study and achieve the learning objectives of this lesson. It will also help you to view the video, which defines and demonstrates important concepts and skills as they relate to everyday life.

Learning Objective	Textbook	Telecourse Student Guide	Background Notes
1. Identify the three broad categories of meteorites and the three types of stony meteorites.	pp. 413–414	Matching Activities: 1, 2, 3, 4, 8, 12; Completion Activities: 1; Self-Test Questions: 1, 2.	
2. Describe the possible origins of the different types of meteorites.	pp. 414–416	Matching Activities: 12, 13; Self-Test Questions: 3, 4.	
3. Describe the characteristics of meteors and meteor showers, and explain how they relate to cometary orbits.	pp. 414–415	Matching Activities: 6, 10, 11; Completion Activities: 2; Self-Test Questions: 5, 6.	
4. Describe the physical properties and orbital characteristics of asteroids.	pp. 416–418	Self-Test Questions: 7, 8.	
5. Explain theories about the origin of asteroids.	p. 419	Self-Test Questions: 9, 10.	
6. Identify and describe the major parts of a comet and explain the effect of solar wind on the tail of a comet.	pp. 420–424	Matching Activities: 5; Completion Activities: 3; Self-Test Questions: 11, 12.	
7. Explain the significance of the Oort cloud and Kuiper belt theories to understanding the origin of comets.	pp. 421–424	Matching Activities: 7, 9; Self-Test Questions: 13, 14.	

Learning Objective	Textbook	Telecourse Student Guide	Background Notes
8. Describe the significant historical observations of Comet Halley, including the 1910 apparition, and the important discoveries made by spacecraft during the 1986 apparition.	pp. 420–421	Self-Test Questions: 15, 16, 17.	Background Notes 25A: Comet Halley—History
9. Explain the possible effects of an asteroid/comet collision with Earth and describe the impact of Comet Shoemaker-Levy 9 on Jupiter.	pp. 424–427	Self-Test Questions: 18, 19.	Background Notes 25B: Comet Shoemaker-Levy 9; Background Notes 25C: Interplanetary Debris Impacts.

26

Life on Other Worlds

Assignments

For the most effective study of this lesson, we suggest that you complete the assignments in the sequence listed below.

Before Viewing the Video Program

- Read the Overview and Learning Objectives for this lesson. Use the Learning Objectives to guide your reading, viewing, and thinking.
- Read Chapter 20, "Life on Other Worlds," pages 430–450, and the "Afterword," pages 451–452, in the *Horizons* textbook.
- Review Chapter 16, section 16-1, "The Great Chain of Origins," pages 323–329.
- Read Background Notes 26A, "Europa: An Oasis for Life?" and 26B, "The Search for Other Planets," following Lesson 26 in this guide.
- Read the Viewing Notes in this lesson.

View the "Life on Other Worlds" Video Program

After Viewing the Video Program

- Briefly note your answers to questions listed at the end of the Viewing Notes.
- Review all reading assignments for this lesson, especially the Chapter 20 summary on page 449 in *Horizons*, Background Notes 26A and 26B for Unit IV, and the Viewing Notes in this lesson.
- Respond to the challenges presented by each Critical Inquiry in the reading assignment and write brief answers to all the review questions at the end of Chapter 20 in *Horizons* to be certain you understand the text material.
- Complete the Review Activities in this guide to reinforce your understanding of important terms and concepts. Check your answers with the Answer Key and review when necessary.

- Take the Self-Test in this guide to measure your achievement of the Objectives. Check your answers with the Answer Key and review when necessary.

- Complete the Using What You've Learned activities and any other activities and projects assigned by your instructor.

Overview

The search for extraterrestrial life is perhaps the greatest challenge presented to science. The search requires the cooperation of several disciplines. Biology is involved because that science defines and describes life as we know it and identifies conditions necessary for life to evolve and be sustained. Astronomy identifies the types of stars that are likely candidates to harbor planets that could support life, and astronomers search for planets around those stars. Engineers and other scientists design and build the radio telescopes and computers that survey the sky and search for signals from extraterrestrial civilizations.

Today, most astronomers think the probability of other stars having planetary systems is very strong. With 100 billion stars in the Milky Way Galaxy alone, it is almost inconceivable that none would have planets. Of course, the existence of planets does not mean that they have life on them. This lesson explores how astronomers are conducting the search for planets and extraterrestrial intelligence and examines the physical basis of life and how life evolved on Earth.

In the video program, astronomers explain why O, B, A, and M type stars are poor candidates for life-harboring planets. By using our solar system as an example, the experts examine the region around our sun that permits life-bearing planets, consider where the existence of liquid water is possible, and explain why water is so necessary.

Although it is reasonably easy to describe the types of stars that are likely to have life-bearing planets in orbit, detecting a planet around another star is not easy. A planet would be millions of times fainter than the star it orbits and would be overcome by the starlight if astronomers tried to observe it directly. Detecting a planet as the result of its gravitational influence on the star would also be difficult, because a planet would be at least 100,000,000 times less massive than the star. In spite of these obstacles, efforts are underway to detect planets around other stars.

The most likely approach to successfully detecting an extraterrestrial civilization, however, is to allow radio telescopes and computers to scan the sky, "listening" for signals that would have the signature of an intelligent life-form. Even that approach has problems. The radio wavelength is extremely broad, so to what frequencies do we tune our instruments? What messages do we listen for? How do we confirm a positive signal? All these questions have been taken into account by the various projects that fall under the general heading of SETI, the search for extraterrestrial intelligence. These searches are scanning the "magic frequencies" and listening to hundreds of stars each day, hoping to receive some positive indication that planets other than Earth harbor intelligent civilizations.

The likelihood of life existing elsewhere in the galaxy is very strong. The abundance of chemicals needed to support and form life is surprising, and these chemicals are present in the star-forming regions of the galaxy. The rules so far understood about the nature of life are so plausible that it seems inconceivable that these rules would not be in force elsewhere and lead to the development of life. The unknown factor is whether that life has evolved into an intelligence capable of communication. Finally, if ever we do contact extraterrestrial civilizations, we must have an answer for the question, "who will speak for Earth and what would our planet say?"

Learning Objectives

When you have completed all assignments for this lesson, you should be able to:

1. Explain the criteria scientists use to define life. *HORIZONS* TEXTBOOK PAGES 431–432.

2. Explain the nature of DNA and RNA and how these molecules help to define life. *HORIZONS* TEXTBOOK PAGES 432–435.

3. Describe the current understanding of how life originated on Earth and the significance of the Miller experiment. *HORIZONS* TEXTBOOK PAGES 436–441.

4. Describe the characteristics of a star and its solar system that would provide the physical conditions allowing life to evolve. *HORIZONS* TEXTBOOK PAGES 443–444; BACKGROUND NOTES 26A; VIDEO PROGRAM.

5. Describe the efforts of astronomers to detect planets around other stars. *HORIZONS* TEXTBOOK PAGES 323–329 AND 443–444; BACKGROUND NOTES 26B; VIDEO PROGRAM.

6. Explain how an equation is used to determine the number of possible extraterrestrial civilizations in the galaxy. *HORIZONS* TEXTBOOK PAGES 447–448; VIDEO PROGRAM.

7. Describe the recent attempts to communicate with extraterrestrial civilizations and to detect radio transmissions from extraterrestrial civilizations. *HORIZONS* TEXTBOOK PAGES 445–447; VIDEO PROGRAM.

Viewing Notes

This program explores one of humankind's oldest questions: Are we alone in the universe?

The first segment of "Life on Other Worlds" identifies the chemicals of life, and astronomers explain how supernovae supplied calcium, iron, and all the other elements that are the basis of life. The video program then examines the role of environment in allowing life-forms to develop and to exist. Astronomers describe the early environment of our solar system and the role of comets and asteroids in the origin of life. This segment also looks at the environments on Venus and Mars and explains why they cannot sustain life.

In the segment "Extrasolar Planets," astronomers explain why detection of planets around other stars is so difficult. They describe some of the techniques they do use, including coronagraphs, infrared sensors, and observations to detect wobbles in a star's movement and Doppler shifts in the light from a star. Astronomers also describe the concept of the habitability zone, the area around a sun where water could exist in liquid form, and explain why only certain types of stars are likely candidates for hosting planets with life on them.

The last segment begins with an explanation of an equation (developed by astronomer Frank Drake), which attempts to estimate the number of technological civilizations in a galaxy. The segment then describes current attempts to search for extraterrestrial intelligence by monitoring radio wavelengths. Astronomers describe the challenges of searching the frequencies spectrum and the use of "magic" frequencies that would be universal.

As you watch the video program, consider the following questions:

1. What are the two most abundant elements in the universe, and where did the heavier elements such as carbon and iron come from and how did they become part of Earth and of us?

2. Why is liquid water an important component in the recipe for life?

3. What role did life itself play in securing its place on planet Earth?

4. What is the habitability zone about a star?

5. Why is it more likely for G and F stars to support life-bearing planets rather than O and A type stars?

6. What are the problems in directly observing or photographing a planet around a distant star?

7. What observations, other than direct detection, can be used to infer the existence of a planet around a distant star?

8. How can observations in the infrared be used to detect the existence of planets around other stars?

9. How does the Drake equation help astronomers determine the potential for extraterrestrial civilizations?

10. What is SETI and how does it use radio waves to listen for extraterrestrial communications?

Review Activities

Matching

Match the terms listed below with the definitions that follow. Check your answers with the Answer Key and review any terms you missed.

_____ 1. DNA   _____ 9. Miller experiment

_____ 2. protein   _____ 10. primordial soup

_____ 3. RNA   _____ 11. chemical evolution

_____ 4. enzyme   _____ 12. stromatolite

_____ 5. amino acid   _____ 13. life zone

_____ 6. natural selection   _____ 14. water hole

_____ 7. mutant   _____ 15. SETI

_____ 8. Cambrian period

a. shorthand reference for the search for extraterrestrial intelligence

b. long carbon-chain molecules that use the information stored in DNA to manufacture complex molecules necessary to the organism

c. interval of the radio spectrum between the 21-centimeter neutral hydrogen radiation and the 18-centimeter OH radiation; likely wavelength to use in the search for extraterrestrial life

d. chemical process that led to the growth of complex molecules on primitive Earth

e. a geological period 0.6 to 0.5 billion years ago during which life on Earth became diverse and complex; rocks from this period contain the oldest identifiable fossils

f. long carbon-chain molecule that records information to govern the biological activity of the organism

g. special protein that controls processes in an organism

h. rich solution of organic molecules in Earth's first oceans

i. region around a star within which a planet can have temperatures that permit the existence of liquid water

j. offspring born with altered DNA

k. complex molecule composed of amino acid units

l. structure produced by communities of blue-green algae or bacteria

m. basic building block of protein

n. experiment that reproduced the conditions under which life may have begun on Earth

o. process by which the most adaptive traits are passed on, allowing the most able to survive

Completion

Fill each blank with the most appropriate term from the list for that paragraph. A term may be used once, more than once, or not at all. Check your answers with the Answer Key and review when necessary. If a question requires two or more answers in succession, they may be in any order, unless the question indicates otherwise.

1. The physical basis for life on Earth is the _____ atom. This atom is capable of forming long, complex, _____ chains that can extract, store, and use _____. These chains are located in the center of the basic unit of life, the _____. Within the _____, these chains take a form that resembles a _____. These complex molecules contain _____ on how the cell _____ and how it _____ as part of the larger organism.

   These complex molecules are called _____, or DNA. The rails of the ladder are made of alternating _____ and _____. Between the rails are _____, which are made of pairs of _____. There are _____ different kinds of _____, the order of which determines how the _____ functions.

   The DNA molecule contains the information necessary to make all the molecules needed by the organism. Segments of the DNA molecule are used in the production of _____, some of which help build the cell wall, while others control body processes such as growth. Those used to control growth are called _____.

   It is not the DNA molecule that actually leaves the cell and begins construction on the _____. The DNA molecule produces a copy of its patterns in a long _____ chain molecule called _____, or RNA. This RNA molecule carries the _____ necessary to use the simple building blocks called _____ and combine them to produce _____.

The RNA molecule acts as a _____, carrying out the patterns established by the _____ molecule.

amino acids

bases

carbon

cell

deoxyribonucleic acid

DNA

eight

energy

enzymes

four

functions

information

messenger

nucleus

phosphates

proteins

reproduces

ribonucleic acid

rungs

stable

sugars

twisted ladder

two

2. Recent research seems to indicate that planets around other stars are probably a likely occurrence. However, planets that might harbor life probably occur much less frequently. The first criteria for a planet to harbor life is that its orbit must be _____. It cannot wander too far or too close to the star. To meet this requirement, the planet would probably orbit a _____ star and not a _____ or multiple-star system.

Furthermore, the lifetime of the star must be very long. It probably requires between 0.5 to 1.0 _____ years for life to evolve on a planet. Therefore, the _____ and _____ type stars are unsuitable to harbor life-sustaining planets because their lifetimes are too _____. Other stars that are poor candidates to harbor life-sustaining planets are _____ stars because they are too _____. Planets around such a star would have to orbit too _____ to the star and may become _____ to the star, keeping the _____ side toward the star. In this situation, one side of the planet would be in _____ darkness, causing the atmosphere and its water to _____. Another problem with the _____ stars is that they are subject to sudden _____. This release of a star's energy with the planet being so _____ might destroy any life that had formed.

The most likely candidates for stars that may have suitable planets to harbor life are the _____, _____, and _____ stars. These stars are _____ and their lives are relatively _____ and stable. They also have a large region of space around them where water can remain in _____ form. This region of space is called a _____, or ecosphere. To summarize, stars that would be suitable to have planets capable of sustaining life are stars that are _____, and of the _____ spectral types: _____, _____, and _____. These stars also have _____ lives and have relatively large _____. At present we know of only one star that meets all these requirements and has produced a planet suitable for sustaining life: our sun.

B	K
billion	life zone(s)
binary	liquid
close	long
constant	M
cool	million
cooler	O
different	same
distant	short
F	single
flares	stable
freeze	tidally locked
G	

3. It seems obvious, because of the great _____ between the stars, that physically _____ to the stars is, at least for now, impossible. Therefore, the only form of _____ we will be able to have with extraterrestrial civilizations will be through the use of _____. These signals have already started leaving Earth in the form of _____ and_____ broadcasts. If any other civilizations pick up these signals, we hope they will still consider us intelligent.

Although scientists may never know if any other civilization has received our signals, either by accident or design, astronomers are

_____ for signals reaching Earth from _____ civilizations. Because the spectrum of _____ is so large, it is necessary to narrow the range of _____ to which telescopes can listen. Astronomers have determined that there are some _____ that will probably be used by extraterrestrial civilizations for this kind of communication. These wavelengths involve _____ and _____ that would be familiar and important to all forms of _____. The range of frequencies lies between the 21-centimeter line of _____ and the 18-centimeter line of the _____ molecule. The interval between these two lines has been called the _____, because the combination of _____ and _____ yields _____ (H_2O). Even if a civilization is not based on _____, it will certainly know, assuming its beings have an intelligence level at least equal to our own, about the _____ line of neutral hydrogen. Current efforts to detect a signal from an _____ intelligence are called _____. Someday we may have an answer from life on another world.

communication	radio
compounds	radio waves
distances	SETI
elements	special (magic) frequencies
extraterrestrial	television
hydrogen	traveling
leaking	21-centimeter
life	water
listening	water hole
neutral hydrogen	wavelengths
OH	

Self-Test

Select the one best answer for each question. Check your answers with the Answer Key and review when necessary.

1. The basic unit of life as we know it is the
   a. nucleus.
   b. cell.
   c. DNA.
   d. RNA.

2. The information that enables the organism to survive and is transferred from parent to offspring is found in
   a. DNA.
   b. the cell membrane.
   c. a twisted ladder.
   d. proteins.

3. The information a cell needs to function is found on the DNA molecule in the form of
   a. multiple combinations of unknown chemicals.
   b. RNA subgroups.
   c. a twisted helix.
   d. four bases and the order in which they appear.

4. The information regarding cell functioning is carried out of the nucleus by
   a. amino acids.
   b. RNA.
   c. proteins.
   d. enzymes.

5. Evidence that life began in the oceans has been found in
   a. surface rocks.
   b. sediments on the ocean floor.
   c. Cambrian period fossils.
   d. the primordial soup.

6. The Miller experiment consisted of various gases, believed to have been present in the early atmosphere of Earth, exposed to an electrical arc. This experiment resulted in

   a. nothing.

   b. carbon dioxide gas and water vapor.

   c. amino acids, fatty acids, and urea.

   d. rain.

7. Hot stars, such as O and A stars, are unlikely to harbor planets suitable for sustaining life because

   a. their stable energy-producing life span is too short.

   b. they are too hot.

   c. they are too cold.

   d. they produce an usually high amount of deadly radiation.

8. The region of space around a star in which water can remain in liquid form is called the

   a. water hole.

   b. temperate zone.

   c. main-sequence stage.

   d. life (or habitability) zone.

9. Scientists think Jupiter's moon Europa might be able to support life because

   a. the Voyager and Galileo spacecraft sent back pictures showing areas of vegetation on its surface.

   b. it gets far more life-giving sunlight than Jupiter's other moons.

   c. there is evidence that below its icy surface there may be liquid water.

   d. its atmosphere contains large quantities of oxygen.

10. Several meteorites found in Antarctica have been determined to come from Mars. The evidence for this conclusion lies in the

    a. analysis of trapped gases in these meteorites with the same abundance as gases found in the Martian atmosphere.

    b. residual magnetism in these meteorites, which is identical to the residual magnetism on Mars itself.

    c. orbital trajectories of these meteorites, which coincide with the position of Mars 16 million years ago.

    d. similarities of organic structures found in rocks on Mars as analyzed by the various spacecraft that have explored the Martian surface.

11. If a large planet orbited a nearby star, astronomers would most likely observe

    a. a faint speck of light very close to the star.

    b. a slight wobble in the star's proper motion.

    c. a red shift in the star's spectrum.

    d. an eclipse as the planet moved in front of the star.

12. As a planet and star orbit about their common center of mass, the center of mass would be observed to

    a. travel across the sky in a straight line.

    b. wobble slightly as the star moved smoothly across the sky.

    c. exhibit blue shift and red shift of the spectral lines as the planet and star orbited.

    d. exhibit only a blue shift in its spectral line as it moved across the sky.

13. The Drake equation is used to

    a. determine the number of stars in the Milky Way Galaxy.

    b. count the number of stars visible in the night sky.

    c. determine the spacing distance of planets in our solar system.

    d. estimate the number of technological civilizations within our galaxy.

14. Of the following factors, the one that is a part of the Drake equation is the

    a. temperature of the star during its main-sequence stage.

    b. number of O and B stars in the galaxy that exhibit the possible existence of orbiting planets.

    c. fraction of the lifetime of a star during which a technological civilization is capable of communicating.

    d. total number of planets in a solar system orbiting about a binary star.

15. One of the most likely wavelengths of radio energy being used to search for extraterrestrial signals is the

    a. scattered blue light of the interstellar medium.

    b. 21-centimeter line of neutral hydrogen.

    c. 3 degrees Kelvin microwave background energy.

    d. Balmer wavelength of neutral hydrogen.

16. The water hole refers to

    a. a range of radio wavelengths (frequencies) that fall between the wavelengths of neutral hydrogen and the OH molecule.

    b. the region around a star in which water can remain in liquid form.

    c. planets that have been confirmed to contain traces of liquid or frozen water.

    d. nebulae of interstellar matter that have water or OH molecules.

Using What You've Learned

1. Make a detailed list of the reasons O, B, A, and M stars are not expected to shelter intelligent civilizations.

2. Make a detailed list of the reasons F, G, and K stars are the most likely to shelter intelligent civilizations.

3. From "The Nearest Stars" in Appendix A, Table A-7, page 457, in *Horizons*, select the stars nearest the sun most likely to shelter civilizations. Be sure to consider spectral type, luminosity class, and whether they are single or multiple star systems. Locate these stars on the SC1 Star Chart that you used in your activities in Lesson 21.

Lesson Review

Lesson 26

Life on Other Worlds

Please Note: Use this matrix to guide your study and achieve the learning objectives of this lesson. It will also help you to view the video, which defines and demonstrates important concepts and skills as they relate to everyday life.

Learning Objective	Textbook	Telecourse Student Guide	Background Notes
1. Explain the criteria scientists use to define life.	pp. 431–432	Matching Activities: 2, 4, 5; Completion Activities: 1; Self-Test Questions: 1, 2.	
2. Explain the nature of DNA and RNA and how these molecules help to define life.	pp. 432–435	Matching Activities: 1, 3, 7; Completion Activities: 1; Self-Test Questions: 3, 4.	
3. Describe the current understanding of how life originated on Earth and the significance of the Miller experiment.	pp. 436–441	Matching Activities: 6, 7, 9, 10, 11, 12; Self-Test Questions: 5, 6.	
4. Describe the characteristics of a star and its solar system that would provide the physical conditions allowing life to evolve.	pp. 443–444	Matching Activities: 13; Completion Activities: 2; Self-Test Questions: 7, 8, 9, 10.	Background Notes 26A: Europa: An Oasis for Life?
5. Describe the efforts of astronomers to detect planets around other stars.	pp. 323–329, 443–444	Self-Test Questions: 11, 12.	Background Notes 26B: The Search for Other Planets?
6. Explain how an equation is used to determine the number of possible extraterrestrial civilizations in the galaxy.	pp. 447–448	Self-Test Questions: 13, 14.	
7. Describe the recent attempts to communicate with extraterrestrial civilizations and to detect radio transmissions from extraterrestrial civilizations.	pp. 445–447	Matching Activities: 14, 15; Completion Activities: 3; Self-Test Questions: 15, 16.	

Unit IV

Background Notes

Radar Astronomy

In current astronomical research, radio astronomy has been invaluable in probing the far reaches of the universe. A specialized type of radio astronomy, **radar astronomy**, has been equally valuable in revealing the secrets of our solar system's planets. Radar astronomy is especially useful for studying planets that are difficult to observe through optical telescopes. Mercury, for example, is too close to the sun for optical study of its surface, and Venus is hidden behind a thick cloud cover.

Radar stands for **ra**dio **d**etecting **an**d **r**anging. In radar astronomy, transmitters send artificially produced radio waves toward the object under study. Analysis of the reflected signal can provide information about a planet's distance from Earth, its rotation rate, and its surface features. To date, the best-known applications of radar in astronomy have been mapping of the surface of Venus and measurement of the rotation rates of Mercury and Venus.

A planet's distance is determined by measuring the time it takes the radio signal to travel to the planet and return to a receiving dish on Earth. If the returning radio "echo" is Doppler-shifted in frequency or wavelength, the speed of the planet in orbit around the sun can be deduced. In the case of Mercury and Venus, this information was used to determine the value of the astronomical unit in miles or kilometers.

Because the planet is rotating, the reflected signal (the original signal was a single frequency or wavelength) has a spread of wavelengths. This spread in wavelength is produced by the radio signal being reflected back from different parts of the planet. A portion of the radio signal is reflected from the part of the planet rotating toward Earth, which causes the signal to be slightly Doppler-shifted toward higher frequencies, or shorter wavelengths. A portion of the radio signal is also reflected off the part of the planet rotating away from Earth, which causes the reflected signal to be Doppler-shifted toward lower frequencies, or longer wavelengths. The spread in wavelengths is used, along with measurements of speed and diameter, to determine the rotation rate of the planet.

Since the planet is spherical, but not perfectly smooth, the reflected signal is not detected back at Earth or in a spacecraft all at once. Because tall mountains and the bottoms of craters are different distances from the transmitter, the reflected signal from these surface features reaches the receiver at different times. By keeping track of the time between signal transmission and reception, as well as any Doppler shift in the signal, a unique point on the planet's surface can be mapped. A computer then compiles all the points together to reveal a picture of the surface of a planet.

Although radio waves are longer than light waves and cannot resolve the smallest of surface features, the Magellan spacecraft was able to create images of features on Venus as small as 100 meters wide, a great improvement over the 100-kilometer resolution capability of the first radar-astronomy images. Radar astronomers are working on methods to improve the resolution to capture even smaller details.

Additional information about the surface of the planet can be acquired by analyzing the intensity of the reflected radar signal. Because radio waves can penetrate the surface of the planet, the material in the surface affects the signal's intensity, and the reflection provides information about the composition and texture of the surface. This type of analysis detected the water ice at the poles of Mercury. This ice, located at the bottom of craters that are permanently hidden from sunlight, is also covered by dark insulating material and cannot be observed in optical wavelengths. Radar signal analysis indicates the bright polar regions could only be produced by the existence of water ice.

You should note that the colors in the radar images you see in the video programs and reproduced in the textbook are artificially added and do not represent the colors one would see on the surface of the planet. Because the human eye is not sensitive to radio wavelength radiation, any image created using radar must be translated, through the use of "false" colors, into an image that the human eye can see. For instance, colors in the Pioneer radar images identify different levels of elevation. The Magellan images of Venus are similar in color to the images returned by the Soviet Venera spacecraft, which photographed the surface in natural light.

BACKGROUND NOTES 21B

Synchronous Rotation

If you were to stand on the surface of the near side of the moon looking toward Earth, you would begin to notice something very strange. Earth would continue to remain high in the sky throughout the entire time that the moon orbited Earth. Earth would never set. Earth appears nearly "fixed" in the sky because the rate at which the moon rotates on its axis is equal to the length of its orbital period around Earth. This phenomenon is referred to as **synchronous rotation**.

In the video program, Bruce Murray, a planetary scientist at California Institute of Technology, explains that the moon is asymmetrical, not perfectly spherical or uniformly dense. This asymmetry causes Earth to exert tidal forces (differences in the gravitational force due to different distances from Earth). These forces eventually slowed the moon's rotation rate until it became locked onto Earth, and its rotation rate became equal to its revolution period.

Mercury is also locked into a peculiar rotation-revolution relationship with the sun. Mercury's orbit is highly eccentric (only the orbit of Pluto is more elliptical). Observations and orbit calculations led astronomers to believe that one side of Mercury faced the sun at all times because of tidal locking, similar to what is in effect with the moon and Earth.

In 1965, however, radar studies using the Arecibo telescope showed that Mercury has a rotation period of 59 days as compared to an orbital period of 88 days. These figures mean that Mercury rotates three times on its axis as it orbits the sun twice. On Mercury, one would observe the sun crossing the meridian (noon) once every 176 days. The length of a solar day on Mercury is equivalent to 176 Earth days. The length of Mercury's day creates some difficulty in studying the surface of the entire planet, even from space. For instance, the Mariner spacecraft flew by Mercury three times and, each time, the same side of Mercury was facing the spacecraft.

B A C K G R O U N D   N O T E S   2 2

Plate Tectonics on Mars?

Plate tectonics is one of the most important geologic forces shaping (reshaping) Earth's surface. Evidence for the motion of crustal plates can be found along the mid-ocean ridges in the Atlantic and Pacific Oceans. As magma (molten rock) rises up and solidifies, the polarity (orientation of the North and South Poles) of Earth's magnetic field is frozen into place. Over time, Earth's magnetic field reverses and new magma solidifies, creating a series of parallel bands of fossilized magnetism on either side of the mid-ocean ridges (See page 340, *Horizons*). As the seafloor spreads and the crustal plates move apart, these bands of fossil magnetism also grow farther apart. Dating the magnetic bands determines the length of time the seafloor spreading has been under way.

The Mars Global Surveyor spacecraft has discovered a similar pattern of bands. As the Mars Global Surveyor, orbiting the planet in a highly elliptical orbit, dipped below the planet's ionosphere, the spacecraft's magnetometer was able to measure subtle differences in the magnetism of a small region of the planet's surface. The pattern of fossilized magnetism showing bands of alternating magnetic polarity is consistent with seafloor spreading observed at the mid-ocean ridges on Earth. The bands are oriented in the east-west direction and are about 100 miles wide and 600 miles long, making them wider than those on Earth. One possible explanation for this difference is that the Martian crust spread more quickly than Earth's crust. A second explanation involves the frequency of magnetic field reversal. The Martian magnetic field may reverse less frequently than Earth's.

The analysis of these magnetic stripes is complicated by the fact that Mars no longer has a global magnetic field and that the evidence of seafloor spreading is very localized. Unlike Earth, where the mid-ocean ridge and the magnetic striping stretches from Iceland to Antarctica, the evidence for seafloor spreading can be found in only two areas. The Martian dynamo that produced the global magnetic field on Mars ceased to exist early in

Martian history. The planet, being half the size of Earth, cooled very quickly, and the molten core solidified and stopped producing the magnetic field. Asteroid impacts and major volcanic eruptions in the Northern Hemisphere (as evidenced by few craters and relatively smooth terrain) destroyed most of the fossilized magnetism that may have supported the conclusions being drawn.

There is still one more piece of evidence missing. Seafloor spreading on Earth is demonstrated by a symmetry of magnetic stripes. The pattern of magnetic field reversal is found on both sides of the mid-ocean ridges. On Mars, only one side of the pattern has been discovered. The point of symmetry has yet to be determined.

The Galileo Probe into the Atmosphere of Jupiter

Traveling at a speed of 170,700 km/hour (106,000 mph), the Galileo Probe entered the atmosphere of Jupiter on December 7, 1995. After entering the atmosphere, the probe continued to transmit data for 57.6 minutes or to a depth of about 200 km (125 miles) below the visible cloud tops. At this point, the communication system failed due to the high temperatures. The probe did not carry a camera. Instead, it had seven scientific instruments, one designed to take measurements of the radiation belts of Jupiter prior to entry and the others designed to take measurements of the properties of Jupiter's atmosphere. A brief summary of the findings of these instruments follows.

The *Energetic Particle Instrument* (EPI) studied the radiation belts of Jupiter prior to the probe's entry into the Jovian atmosphere. The EPI discovered an inner belt with a radiation intensity 10 times greater than the Van Allen radiation belts circling Earth. Further study of the data will increase our understanding of Jupiter's magnetosphere. In turn, understanding more about strong magnetic fields, will help scientists understand the environment around pulsars, stars, and galaxies.

The *Atmosphere Structure Instrument* (ASI) measured temperature, pressure, and vertical winds in Jupiter's atmosphere. The pressure increased from the uppermost regions to a maximum of about 24 times Earth's atmospheric pressure. This is equivalent to swimming at a depth of about 230 meters (750 feet) in the ocean. In addition to an increase in pressure, temperature also increased from about $-200\ ^{0}$F at the cloud tops to $+305\ ^{0}$F at the position of the probe. Two conclusions can be drawn from these measurements and the variation of temperature and pressure with depth. The deep atmosphere is dryer than expected and is influenced by convection. Furthermore, the higher than expected temperatures require an additional source of heating and not merely sunlight. This additional source of heat comes from the interior of the planet.

Two instruments were designed to measure densities of the clouds in the vicinity of the probe entry site. The *Nephelometer* instrument (NEP), designed to measure nearby cloud particles, and the *Net Flux Radiometer* (NFR), designed to measure distant cloud particles, provided scientists with a surprise. No thick dense clouds were found in the vicinity of the probe entry site. This result is contrary to expectations based on telescopic and flyby spacecraft observations of the planet. Only small concentrations of cloud and haze material were found along the entire descent path. The conclusions drawn from these two cloud detecting instruments is that the clouds of Jupiter are patchy and that the probe went through a unique are relatively clear of clouds.

The *Doppler Wind Experiment* measured the speed of the Jovian winds at various depths. The winds blow at a speed of 700 km/hour (435 mph) and seem to be independent of depth. The speeds of winds at the top of Jupiter's clouds are of similar strength as measured by the Hubble Space Telescope. The conclusion drawn from this observation is that the winds of Jupiter are powered not by sunlight and water condensation as they are on Earth, but by heat escaping from Jupiter's interior.

The *Lightning and Radio Emission Detector* measured the presence of lightning discharges by searching for optical flashes and radio wave emission. No optical flashes were observed and the radio emissions detected indicated that the lightning occurred at a great distance from the probe, possibly as much as one Earth radii. Although the lightning bolts on Jupiter are much stronger than those on Earth, their frequency is 3 to 10 times less than on Earth. These results are different than what was expected and therefore ideas of water cloud distribution, precipitation, and heat loss from Jupiter may need revision by scientists.

Jupiter, the largest planet in the solar system, has the strongest gravitational grip on the material of which it is comprised. Composed mostly of hydrogen and helium, Jupiter also contains oxygen, carbon, neon, and nitrogen—the next four most abundant elements in the universe. The detailed percentages of abundance of these gases will help to provide clues to the process of planetary formation and evolution. The *Neutral Mass Spectrometer* (NMS) and *Helium Abundance Detector* found the composition of Jupiter to be what was expected although it was surprising at other times. The abundance of hydrogen and helium was similar to the Sun, but oxygen is much less abundant than in the Sun's atmosphere. The lack of oxygen, as measured by the water vapor content in the atmosphere, indicates that Jupiter has a very dry atmosphere. The abundance of carbon (methane gas) and sulfur (hydrogen sulfide gas) occur at levels greater than solar amounts. The noble gas neon is highly depleted and the percentage of nitrogen is as yet undetermined. There is little evidence of organic molecules in the atmosphere of Jupiter.

Scientists had expected to find oxygen in greater abundance than solar levels because of the number of cometary impacts that must have occurred during the 4.5-billion-year history of the planet. Helium abundance was expected to be lower due to the internal evolution of the planet. The dryer than expected atmosphere of Jupiter will require a revision of the role meteorology plays in a planet's history. These discoveries will provide scientists with much information to study and some significant questions to be answered.

Comet Halley—History

The first recorded observation of the celestial object we know as Comet Halley was in 240 B.C. Throughout much of Western civilization and history, comets, especially Comet Halley, have been linked to catastrophic events. This association may have started with Aristotle, who noted the appearance of comets before and after severe storms. Although the storms and the comets were not related, Aristotle's speculations, supported by his reputation and stature, may have given rise to superstitions that connected comets with war, disease, pestilence, famine, and especially the deaths of kings and nobility.

For instance, comets, have often been associated with the destruction of city states and the overthrow of kings. In the first century A.D., Josephus, the Jewish historian, described an omen that foretold the destruction of Jerusalem: a bright comet that "hung like a sword." Comet Halley was observed in A.D. 66, and the Romans attacked Jerusalem in A.D. 69. One should, however, be careful about confusing a prediction with a connection. Josephus wrote about the comet and the fall of Jerusalem after the fact. Other historically significant events that have been associated with apparitions of Comet Halley include the defeat of Attila the Hun in 451, a plague that spread across Europe in 530, and the eruption of Mt. Vesuvius in 684.

The most famous apparition of Comet Halley in European history was associated with the Battle of Hastings in 1066. This battle occurred when Duke William of Normandy invaded and conquered England. William may have observed Comet Halley in the western sky in the spring of 1066, and this observation may have motivated him to invade England later that fall. A depiction of the Battle of Hastings and the return of Comet Halley has been embroidered into the Bayeux Tapestry. This 231-foot-long medieval tapestry consists of 72 scenes that illustrate the Norman conquest of England.

Although barely mentioned in European records, the most impressive and spectacular apparition of Comet Halley occurred in 837 and was carefully recorded by Chinese astronomers. In March of that year, the tail of the comet was more than 20 times the diameter of the full moon. Within three weeks, the tail stretched across nearly one-third of the entire sky and had split into two sections. Modern calculations have since determined that the comet the Chinese observed came within 5 million kilometers of Earth. During its closest approach, Comet Halley traversed one-fourth of the entire sky in a period of one day. At its maximum, it was nearly as bright as Venus, and the tail lengthened to 76°, or nearly half the sky. It was during this apparition that Chinese astronomers concluded that the tail of a comet always pointed away from the sun, a fact that would take Western astronomers nearly 700 years to rediscover.

Although the most spectacular apparition occurred in 837, the three most scientifically significant apparitions occurred in 1531, 1607, and 1682. Those apparitions were neither the brightest nor the closest, but they were the three recorded encounters that Edmund Halley used to calculate the orbit of the comet and predict its return in 1758. These calculations, based on Isaac Newton's law of universal gravitation, were the first demonstration that the laws of physics discovered on Earth were applicable to objects millions of kilometers away.

Superstitions about Comet Halley resurfaced at the time of its appearance early in the twentieth century. Even though astronomers had identified some of the gases in the tail of the comet and explained that the tail was very thin and would cause no problems on Earth, the announcement that Earth would pass through the tail of Comet Halley in 1910 caused many people to panic and a number of people to go into business. Comet pills were very popular in the southeastern United States, and breathing masks were also in demand. It was also rumored that if one stood in a bucket of rainwater while the comet was overhead, the gases in the tail would be dissolved by the water and the individual would remain unharmed. Even the American author Mark Twain used Comet Halley to make a prediction about his own life. He said in 1909, "I came in with Halley's Comet in 1835. It is coming again next year, and I expect to go out with it. It will be the greatest disappointment of my life if I don't go out with Halley's Comet. . . . Oh! I am looking forward to that!" Mark Twain did go out with Comet Halley in 1910. Superstitions are revived with each apparition of Comet Halley and will probably continue into the future. The next apparition of Comet Halley will be in July of 2061.

Comet Shoemaker-Levy 9

In March of 1993 astronomers Gene and Carolyn Shoemaker and David H. Levy announced the discovery of a comet. Although many comets are discovered each year by professional as well as amateur astronomers, this comet was special. First, it was one of the rare comets that was in orbit about Jupiter instead of the sun. Orbital calculations showed that the comet had been captured by Jupiter's gravitational influence more than a century ago. Second, Comet Shoemaker-Levy 9, as it was named, had broken into 20 fragments. Orbital calculations again showed that the disintegration of the comet took place during its closest approach to Jupiter in July of 1992. It was clear to astronomers that the comet was not strongly put together because the tidal forces exerted by Jupiter on the comet were not very great. Finally, and most important, Comet Shoemaker-Levy 9 was on a collision course with Jupiter itself, scheduled to impact the planet in July of 1994.

After spreading out over the two years from breakup to impact, the 20 fragments approached Jupiter in a thin line. In fact, astronomers calculated that it would take five and a half days for all the fragments to strike the planet. In predicting the consequences of the impacts, astronomers faced several problems. The first, and probably the most difficult to resolve, was to estimate the size of each piece. At a distance of 480 million miles from Earth, the comet had a halo of gas and dust around it that made it difficult to determine the size of each piece. Estimates of the size of the pieces ranged from 200 feet (60 meters) to 2 miles (5 kilometers). The second problem was to determine how deeply into the Jovian atmosphere the fragments would penetrate. The pieces, traveling at 130,000 miles an hour, were either going to explode upon impact with the atmosphere or penetrate deep into the clouds, causing material from Jupiter to erupt into a giant plume.

When the fragments, labeled A through W, finally struck Jupiter, they released more energy than if all the nuclear weapons on Earth were to be detonated at once. The first and smallest fragment produced a fireball 4,000 kilometers across and a plume 1,000 kilometers high. The energy released was equal to more than a million times the energy released by the A-bomb dropped on Hiroshima, Japan, at the end of World War II. Although the impact sites were on the far side of Jupiter, the planet's rapid rotation brought the impact sites into view of Earth telescopes within a half hour. Fortunately, however, the Galileo spacecraft was in a position to photograph the moment of impact for some of the fragments.

The largest fragment, labeled G, was estimated to be 5 kilometers (2 miles) across. It struck Jupiter with a force equal to 6 million megatons of TNT. The blast released energy equal to that of 6 million million tons of dynamite, all in just a few minutes. In comparison, the United States was estimated to have 20,000 megatons and the Soviet Union 60,000 megatons of explosive power during the height of the Cold War. Although the blast from fragment G created a plume of material 2,000 kilometers high, there was no evidence, from spectroscopic studies, of water in the plume. Astronomers concluded that the explosion was relatively shallow and no material from deep within Jupiter was ejected. The impact of G and the other largest pieces also produced dark spots thousands of kilometers across that were easily viewed from Earth-based telescopes. Preliminary results indicate that these dark spots are composed of material from the comet itself and not material from Jupiter's interior.

Observatories throughout the world focused on this display. Viewing the Comet Shoemaker-Levy 9 impacts in visible as well as infrared light, astronomers witnessed an event that has been repeated countless times since the formation of the solar system.

Interplanetary Debris Impacts

Comet Shoemaker-Levy 9 was a spectacular instance of interplanetary debris hitting a planet. Such impacts have occurred throughout the solar system, including Earth. On Earth, most evidence of impacts has been obliterated through erosion and the effects of plate tectonics. However, some evidence of such impacts from either comets or asteroids has been found in several places, including near the Yucatán Peninsula in Mexico and in Siberia. Also, military records contain data about possible impacts of space debris on Earth's surface or in Earth's atmosphere.

Extinction of the Dinosaurs

In 1980 Luis and Walter Alvarez of the University of California at Berkeley proposed that a large meteorite or cometary impact, occurring at the end of the Cretaceous period, was responsible for the mass extinction of the dinosaurs and many other species. They based their conclusion on the existence of the element iridium in a layer of Earth's crust dating 65 million years ago. Iridium is a rare element on Earth, but thousands of times more abundant in primitive meteorites. This iridium-rich layer has been found at more than 100 locations throughout the world.

If the Alvarezes are correct, a 10-kilometer object, either an asteroid or a comet traveling at a speed of 15 kilometers a second, collided with Earth. This object was three times larger than the largest fragment of Shoemaker-Levy 9. The impact ejected millions of tons of dust into the air, eliminating or significantly reducing the amount of light reaching the surface of Earth. Fires, noxious fumes, and acid rain would also have accompanied the impact. In addition to the biological consequences, the asteroid/comet would also have left a crater 150 to 200 kilometers wide. Since the Alvarezes first proposed their hypothesis, many scientists searched for that crater.

The crater may have been discovered in 1978 by the Mexican national oil company, Pemex. Geologists conducting a survey of magnetic anomalies off the coast of the state of Yucatán discovered a circular feature resembling an impact crater. The crater, buried under water, rock, and soil, was very difficult to decipher. (Details of the discovery process of this crater can be found in an article in the July 1991 issue of *Astronomy*.) Although other craters are being considered as possible sites for an impact causing the mass extinctions 65 million years ago, many planetary scientists accept the hypothesis that such an impact did end the reign of the dinosaurs. The question concerns the possibilities of a smaller or similar impact occurring on Earth in the near future.

Siberian Explosion and Other Evidence

In 1908 a large explosion occurred over the Tunguska region of Siberia. Witnesses described a fiery tail striking Earth. A man, 60 miles away, was knocked down, and shock waves were recorded in England. In Siberia a forest roughly the size of Rhode Island was leveled. (Evaluations of the damage done and the fact that no crater exists indicate that the object, either a comet or an asteroid, exploded 4 miles above Earth's surface.)

In addition to the evidence from Siberia, the U.S. military has kept classified records of high-altitude impacts and collisions with Earth's surface. These records show that, on the average, there are eight explosions a year in the upper atmosphere caused by meteors. These explosions are equal in magnitude to a small atomic bomb. One explosion, believed to be an asteroid, occurred in the Pacific in 1978 and was equal to 100 kilotons of TNT. In 1991 an asteroid, estimated to be 30 feet in diameter, passed between Earth and the moon. A 2,200-pound meteorite exploded over the city of Mbale, Uganda, in 1992. Fragments damaged a train station and cotton factory. One boy was struck in the head by a fragment. He survived.

Because of its larger size, Earth has been struck more often by asteroids and meteorites than has the moon. It is anticipated that Earth will be struck again in the future. Planetary scientists are searching the skies to try to predict if Earth is in any immediate danger. Others are trying to decide what to do in case an object is discovered to be on a collision course. The search for objects will continue, along with the search for a solution to the problem.

Europa: An Oasis for Life?

It has long been the wish of human beings to answer the question "Are we alone?" in the negative. For most, that answer would mean that extraterrestrial life exists, that this life is intelligent, and that it has the capability and desire to communicate with the intelligent life on Earth. At present, there is no conclusive evidence that an extraterrestrial intelligence has made contact with Earth, and that much-desired answer may never come. However, the question "Is their life in the universe (even elsewhere in our solar system) beyond the surface of Earth?" may be answered in the positive within the next ten years. Evidence for the existence of extraterrestrial life is contained in photographs taken of Jupiter's moon Europa by the Galileo spacecraft.

When the Voyager spacecraft sent back the first images of Europa in the late 1970s, planetary scientists deduced that the surface of Europa most probably was ice and the interior liquid water. This conclusion was based on the very smooth appearance of the moon and the fact that the surface was criss-crossed with dark lines (see Figure 18-9 in *Horizons*). These findings were important because biologists consistently asserted that an environment that could support primitive life must gave liquid water in it. Liquid water permits the movement of chemicals and organisms to allow for the intake of nutrients and for reproduction. The conclusions described were only suppositions. Voyager was unable to resolve the lines and other surface features into indisputable evidence for the existence of ice and liquid water. That evidence was forthcoming with the April

1997 photographs from the Galileo spacecraft, taken at a distance of 2,075 miles above Europa's surface.

The first thing one notices when viewing the images taken by Galileo is that the surface of Europa is not smooth. [See *The Planetary Report* 17, no. 2 (March/April 1997), or visit the following site—http://www.jpl.nasa.gov/galileo/index.html.] The most prominent features are a multitude of parallel ridges and grooves. Close examination reveals that these ridges have been offset by the movement of Europa's crust or obliterated by what appears to be ice floes. These ridges and grooves are believed to be the result of water, combined with other materials, oozing up through long parallel cracks. These cracks are analogous to volcanic vents found on the surface of Earth. Furthermore, the ridges form on both sides of these "volcanic vents" mirroring processes at Earth's midoceanic ridges, where molten rock reaches the surface and forms new crust. One last feature supporting the conclusion that there is a liquid ocean beneath Europa's crust was the discovery of icebergs. Since the icebergs show only 10 percent of their size above water (some 300–600 feet) scientists have concluded that this icy crust is probably no more than a mile or so thick.

The observed activity on the surface of Europa is fairly recent. There are practically no craters on the surface and those that are present are relatively small. This means that Europa must have some internal heat source maintaining the liquid water that is the cause of all these features. (This is surprising, since Europa is only

2,000 miles across and should have solidified long ago.) That heat source is the same tidal flexing that produces the heat needed to power the volcanoes on Jupiter's moon Io. Jupiter and its other Galilean satellites are in a tug-of-war with Europa. The tugging and pulling by these objects squeezes Europa like a sponge, generating heat that melts the mostly icy world.

The evidence for liquid water below the crust of Europa is convincing. The discovery of brown stains on the surface of this moon is encouraging. They may be hydrogen cyanide or other life-related compounds. "If this is true," says Richard Terrile, a planetary scientist at the Jet Propulsion Laboratory in Pasadena, California, "we have organic chemicals mixed into a bath of water. That's a recipe for life." The definitive answer may come within the next ten years. NASA is planning more missions to the planetary satellites of Jupiter. If radar studies by one of these spacecraft confirms the existence of water beneath the surface, another vehicle will be sent to set down on the surface of Europa and release a probe to melt through the icy crust and search for life in the oceans below.

If life can be confirmed to exist elsewhere in the solar system, then is it safe to conclude that life will form whenever the environment and conditions for life exist? If life is as versatile and abundant in our galaxy (in the universe) as many scientists think, should we still consider ourselves alone? The chemistry of life is hardy. What is the probability that simpler life-forms will evolve (may already have evolved) into an intelligence capable of interstellar communications, or better, interstellar travel?

The Search for Other Planets

Detection of planets around other stars is extremely difficult. At present, it is impossible to see directly a planet around another star. No telescopes on Earth or in orbit above Earth can separate a planet and its star into two images. The star is too bright and the planet too faint. Astronomers at some observatories, however, use a coronagraph to block out the light of the star, producing an artificial eclipse. By blocking out the bright light of the star, astronomers may be able to observe the fainter planet.

Other astronomers are trying to detect a planet in orbit about a star by looking for evidence of its gravitational effects. If no planets are in orbit about a star, the star will move across the sky in a straight line. In other words, the star's proper motion would be in a straight line. However, if a planet does orbit a star, the planet and star orbit around a common point called the center of mass. This point is the same center that two stars in a binary system orbit as described in Chapter 8, pages 142–143 (and Figure 8-11) in the textbook. (The center of mass would be much closer to the star than to the planet, because the star is much more massive.) In this case, the center of mass of the planet-star system would move across the sky in a straight line, and the star's path would appear to wobble slightly as it moved across the sky. The planet and star tug on each other gravitationally, and this tug-of-war is revealed in the wobbling motion of the star. Such an observation would require decades of very careful measurement to confirm the existence of a planet.

Another possible method of discovering a planet around another star is to look for a Doppler shift in the light of the star. This shift in wavelength of the light would be due to the wobbling the planet would cause as the star and planet orbited the common center of mass. This effect is similar to observations made of spectroscopic binaries as described in Chapter 8, pages 144–146 (and Figure 8-14) in the textbook. The effect is more difficult to detect because the wobble again is not as extreme as in a binary star system because a planet has so much less mass than a star.

Finally, some astronomers are searching for planets with instruments that detect infrared energy. Stars are much brighter in visible light than planets. They may be as much as a billion times brighter. In the infrared, however, stars are only 10,000 to 100,000 times brighter than planets. Infrared measurements may prove to be a very useful approach to detecting the presence of planets around other stars.

As of the printing of this student guide, more than 100 extrasolar planets have been detected.

Answer Key

Lesson 1

Matching

1. b (Objective 1; page 4)
2. a (Objective 3; page 4)
3. f (Objective 4; page 4)
4. d (Objective 4; page 5)
5. e (Objective 1; page 6)
6. c (Objective 1; page 6)

Completion

1. star, Milky Way, stars, cluster, supercluster, filaments, universe, light-years, time, millions, billions (Objectives 1 and 4; pages 5–7)

2. Earth, star, sun, Earth, sun, astronomical unit, sun, astronomical units, sun, astronomical units, time, 8 minutes, 4 hours (Objective 1; pages 3–5)

   thousands, years, light-year, distance, year, 63,000 astronomical units, Proxima Centauri, light-years, years (Objective 2; pages 4–5)

Self-Test

1. c (Objective 1; pages 4–7; video program)
2. a (Objective 1; page 6; video program)
3. b (Objective 2; page 5)
4. d (Objective 2; page 5)
5. c (Objective 3; page 5 and Appendix A, pages 455–456)
6. d (Objective 3; page 4 and Appendix A, page 455)
7. c (Objective 4; page 5; video program)
8. a (Objective 4; page 4; video program)
9. b (Objective 5; Background Notes 1)
10. b (Objective 5; Background Notes 1)

Lesson 2

Matching

1. c (Objective 1; page 10)
2. g (Objective 1; page 10)
3. q (Objective 4; page 12)
4. f (Objective 4; page 12)
5. j (Objective 2; page 16)
6. n (Objective 6; page 16)
7. e (Objective 6; page 16)
8. a (Objective 5; page 22)
9. l (Objective 6; page 22)
10. p (Objective 6; page 22)
11. i (Objective 6; page 22)
12. b (Objective 6; page 22)
13. k (Objective 7; pages 14–15)
14. o (Objective 6; page 25)
15. h (Objective 6; page 25)
16. m (Objective 2; page 24)
17. d (Objective 2; page 24)
18. u (Objective 2; page 20)
19. w (Objective 2; page 21)
20. v (Objective 2; page 20)
21. x (Objective 5; page 21)
22. t (Objective 2; page 21)
23. r (Objective 2; page 14)
24. s (Objective 2; page 20)

Completion

1. celestial sphere, celestial equator, north celestial pole, south celestial pole, ecliptic (Objective 2; pages 20–21)

2. axis of rotation, tilted, ecliptic, celestial equator, ecliptic, autumnal, vernal, night, day, equal, autumn, spring, summer solstice, highest, winter solstice, lowest, summer, winter (Objectives 5 and 6; pages 21–23)

3. Hipparchus, six, brightness, magnitude, one, six, brightest, smallest, first, sixth, negative, brightness, distance, magnitudes, apparent visual magnitudes, appear (Objective 4; pages 12–13)

Self-Test

1. c (Objective 1; page 11; video program)
2. d (Objective 1; page 10)
3. a (Objective 2; pages 14, 16; video program)
4. d (Objective 2; page 16; video program)
5. c (Objective 3; video program)
6. b (Objective 3; Background Notes 2A; video program)
7. a (Objective 4; pages 12–13)
8. b (Objective 4; pages 12–13)
9. c (Objective 5; pages 21–22)
10. d (Objective 5; page 22)
11. a (Objective 6; page 22)
12. c (Objective 6; page 22)
13. b (Objective 7; pages 14–15)

14. a (Objective 7; page 15)
15. b (Objective 8; Background Notes 2B; video program)
16. c (Objective 8; Background Notes 2B; video program)

Lesson 3

Matching

1. h (Objective 2; page 29)
2. m (Objective 2; page 29)
3. a (Objective 4; page 27)
4. i (Objective 4; page 27)
5. q (Objective 5; page 30)
6. g (Objective 5; page 30)
7. n (Objective 5; page 30)
8. b (Objective 6; page 31)
9. k (Objective 6; page 32)
10. r (Objective 6; page 32)
11. f (Objective 6; page 32)
12. d (Objective 6; page 32)
13. p (Objective 6; page 33)
14. j (Objective 6; page 33)
15. o (Objective 6; page 33)
16. c (Objectives 5 and 6; page 35)
17. s (Objective 6; page 35)
18. l (Objective 6; page 36)
19. e (Objective 6; page 33)

Completion

1. moon, Earth, umbra, photosphere, corona, chromosphere, prominences (Objective 6; pages 30–33)
2. new, full, tilted, Earth's, Earth's, nodes, eclipse seasons, two, six, elliptical, apogee, smaller, photosphere, ring, annulus, annular (Objectives 5 and 6; pages 30–34)

Self-Test

1. c (Objective 1; page 28; video program)
2. a (Objective 1; page 28; video program)
3. b (Objective 2; page 29; video program)
4. c (Objective 2; page 29; video program)
5. b (Objective 3; page 28)
6. d (Objective 3; page 28; video program)
7. a (Objective 4; pages 26–27; video program)
8. b (Objective 4; page 27; video program)
9. c (Objective 5; page 30; video program)
10. d (Objective 5; page 30; video program)
11. b (Objective 6; page 31; video program)
12. a (Objective 6; pages 34–35; video program)

13. d (Objective 6; pages 35–36; video program)

14. a (Objective 6; pages 33–34; video program)

15. d (Objective 7; pages 37–39)

16. c (Objective 7; pages 37–39)

Lesson 4

Matching

1. f (Objective 2; page 44)
2. c (Objective 2; page 44)
3. i (Objective 2; page 44)
4. h (Objective 3; page 44)
5. b (Objective 2; page 45)
6. n (Objective 2; page 45)
7. k (Objective 2; page 45)
8. a (Objectives 4 and 5; page 47)
9. g (Objective 9; page 55)
10. e (Objective 9; page 14)
11. m (Objective 9; page 55)
12. j (Objective 9; page 55)
13. d (Objective 7; page 54)
14. l (Objective 9; page 49)

Completion

1. geocentric, Earth, stationary, circular, constant, uniform circular, retrograde, epicycles, constant, deferents (Objectives 2 and 3; pages 44–45)

   off center, constant, equant (Objectives 2 and 3; pages 44–45)

2. Tycho Brahe, three, heliocentric, sun (Objective 7; pages 52–56)

   ellipses, sun, equal, equal, perihelion, fastest, slowest, square, cube (Objective 7; pages 52–56)

   empirical, observations, derived (Objective 7; pages 52–56)

Self-Test

1. c (Objective 1; Background Notes 4)
2. a (Objective 1; Background Notes 4)
3. c (Objective 2; page 44)
4. a (Objective 2; page 44)
5. c (Objective 2; page 45; video program)
6. b (Objective 3; page 44)
7. d (Objective 3; page 44; video program)
8. b (Objective 4; page 47)
9. d (Objective 4; pages 47–48; video program)
10. c (Objective 5; page 45; video program)
11. d (Objective 5; page 47; video program)
12. a (Objective 6; pages 50–51; video program)
13. b (Objective 6; page 51; video program)
14. b (Objective 7; page 54)
15. d (Objective 7; page 58)
16. c (Objective 8; page 57; video program)
17. a (Objective 8; pages 56–58; video program)
18. c (Objective 9; page 55)
19. a (Objective 9; page 14)
20. c (Objective 9; page 49)

Lesson 5

Matching

1. d (Objectives 1 and 2; pages 61–62)
2. k (Objectives 1 and 2; pages 61–62)
3. b (Objectives 1 and 2; page 62)
4. n (Objectives 1 and 2; page 61)
5. h (Objective 1; pages 61–62)
6. o (Objective 1; pages 61–62)
7. g (Objective 1; pages 61–62)
8. m (Objective 2; page 62)
9. i (Objective 4; Background Notes 5B)
10. a (Objective 5; Background Notes 5B)
11. l (Objective 6; Background Notes 5B)
12. e (Objective 5; Background Notes 5B)
13. j (Objective 8; Background Notes 5B)
14. c (Objective 8; Background Notes 5B)
15. f (Objective 8; Background Notes 5B)
16. p (Objective 7; Background Notes 5B)

Completion

1. rest, continue, constant, direction, force, change, proportional, change, same, force, equal, force, opposite, action-reaction, universal, proportional, masses, inversely, square (Objectives 1 and 2; pages 60–63)

2. same, speed, light, space, time, light, dilation, slowing, moving, rest, muon, brief, time dilation, muon, slower, energy, energy, mass, light, stars (Objectives 4, 5, and 6; Background Notes 5B)

3. gravity, acceleration, principle of equivalence, impossible, acceleration, gravitational, curvature, four, precesses, bend, slows, red shift, gravity, space-time, light, neutron, decays (Objectives 7 and 8; Background Notes 5B)

Self-Test

1. c (Objective 1; pages 61–62)
2. a (Objective 1; pages 61–62)
3. b (Objective 2; pages 61–62; Background Notes 5A)
4. c (Objective 2; pages 62, 64–65; Background Notes 5A)
5. d (Objective 3; Background Notes 5A; video program)
6. a (Objective 3; Background Notes 5A)
7. b (Objective 4; Background Notes 5B; video program)
8. d (Objective 4; Background Notes 5B; video program)
9. a (Objective 5; Background Notes 5B; video program)
10. c (Objective 5; Background Notes 5B; video program)
11. a (Objective 6; Background Notes 5B; video program)

12. b   (Objective 6; Background Notes 5B; video program)
13. b   (Objective 7; Background Notes 5B; video program)
14. c   (Objective 7; Background Notes 5B; video program)
15. d   (Objective 8; Background Notes 5B; video program)
16. d   (Objective 8; Background Notes 5B; video program)
17. a   (Objective 8; Background Notes 5B; video program)

Lesson 6

Matching

1. d (Objective 1; page 69)
2. e (Objective 1; page 69)
3. c (Objective 1; page 70)
4. f (Objective 1; page 71)
5. l (Objective 2; page 72)
6. p (Objective 2; page 71)
7. k (Objective 2; page 72)
8. n (Objective 2; page 72)
9. o (Objective 2; page 72)
10. m (Objective 2; page 72)
11. h (Objective 2; page 78)
12. j (Objective 2; page 78)
13. i (Objective 2; page 78)
14. g (Objective 2; page 79)
15. a ((Objective 2; page 79)
16. b (Objective 2; page 79)
17. v (Objective 3; page 73)
18. x (Objective 3; page 73)
19. u (Objective 3; page 74)
20. w (Objective 3; page 75)
21. r (Objective 4; page 81)
22. t (Objective 4; page 82)
23. s (Objective 4; page 82)
24. q (Objective 5; page 84)

Completion

1. greater, lesser, gamma, X rays, radio, ultraviolet, infrared (Objective 1; pages 70–71)
2. lens, objective, greater, eyepiece, chromatic (Objective 2; pages 72–73)
3. mirror, upper, secondary mirror, lens, glass (Objective 2; pages 72–73)
4. dish, poorer, interferometer, equal (Objective 5; pages 84–85)

Self-Test

1. a (Objective 1; page 70)
2. b (Objective 1; page 71)
3. d (Objective 2; pages 78–80; video program)
4. b (Objective 2; pages 77, 79)
5. c (Objective 3; pages 73–77; video program)
6. d (Objective 3; pages 75–77)
7. d (Objective 4; page 82)
8. b (Objective 4; pages 82–83)
9. a (Objective 5; pages 84–87; video program)
10. d (Objective 5; page 84; video program)
11. d (Objective 6; page 87)
12. c (Objective 6; page 89)

Lesson 7

Matching

1. j (Objective 1; page 93)
2. v (Objective 1; page 93)
3. x (Objective 1; page 93)
4. w (Objective 1; page 93)
5. u (Objective 1; page 94)
6. e (Objective 1; page 94)
7. o (Objective 1; page 94)
8. b (Objective 1; page 94)
9. n (Objective 1; pages 94–95)
10. f (Objective 1; page 95)
11. l (Objective 1; page 95)
12. k (Objective 1; page 96)
13. r (Objective 1; page 96)
14. q (Objective 1; page 96)
15. c (Objective 2; page 97)
16. d (Objective 2; page 100)
17. i (Objective 2; page 100)
18. y (Objective 2; page 100)
19. z (Objective 2; page 100)
20. g (Objective 2; page 100)
21. a (Objective 2; page 100)
22. m (Objective 3; page 101)
23. s (Objective 3; page 101)
24. p (Objective 4; page 103)
25. t (Objective 5; page 105)
26. h (Objective 5; page 105)

Completion

1. continuous, temperature, bright-line, chemical composition, dark-line, temperature, chemical composition, temperature (Objective 2; pages 99–102)

2. nucleus, electrons, protons, neutrons, protons, neutrons, electrons (Objective 1; pages 93–94)

3. lesser, lesser, longer, lesser, greater, shorter (Objective 3; pages 99–102)

Self-Test

1. c (Objective 1; pages 96–97; video program)
2. a (Objective 1; pages 96–97)
3. b (Objective 2; pages 97–98)
4. c (Objective 2; pages 100–101; video program)
5. b (Objective 3; page 101)
6. b (Objective 3; page 101)
7. d (Objective 4; page 102)
8. c (Objective 4; page 103)
9. c (Objective 5; page 107; video program)
10. c (Objective 5; page 105; video program)

Lesson 8

Matching

1. l (Objective 3; page 125)
2. j (Objective 2; page 120)
3. b (Objective 1; page 113)
4. c (Objective 1; page 114)
5. m (Objective 3; page 125)
6. i (Objective 2; page 120)
7. n (Objective 1; page 113)
8. k (Objective 3; page 125)
9. o (Objective 1; page 113)
10. p (Objective 1; page 116)
11. g (Objective 2; page 119)
12. h (Objective 4; page 119)
13. a (Objective 1; page 112)
14. d (Objective 1; page 116)
15. q (Objective 1; page 113)
16. e (Objective 2; page 117)
17. r (Objective 1; page 114)
18. f (Objective 2; page 119)

Completion

1. photosphere, chromosphere, corona (any order); photosphere; thin; low; solar; chromosphere, corona (either order) (Objective 1; pages 112–115)
2. rotation, faster, higher, faster, interior, differential, magnetic (Objective 2; pages 120–121)

Self-Test

1. a (Objective 1; page 112; video program)
2. d (Objective 1; page 115; video program)
3. a (Objective 1; page 112)
4. c (Objective 2; page 120; video program)
5. d (Objective 2; page 119; video program)
6. b (Objective 2; video program)
7. b (Objective 3; Background Notes 8)
8. b (Objective 3; Background Notes 8; video program)
9. d (Objective 4; page 119; Background Notes 8)
10. b (Objective 4; page 119; Background Notes 8)

Lesson 9

Matching

1. d (Objective 1; page 134)
2. h (Objective 1; page 135)
3. i (Objective 2; page 137)
4. b (Objective 3; page 138)
5. f (Objective 3; page 139)
6. o (Objective 3; page 140)
7. p (Objective 3; page 140)
8. l (Objective 3; page 140)
9. m (Objective 3; page 141)
10. a (Objective 4; page 142)
11. g (Objective 4; page 143)
12. j (Objective 4; page 144)
13. q (Objective 4; page 144)
14. e (Objective 4; page 147)
15. n (Objective 4; page 147)
16. c (Objective 5; page 149)
17. k (Objective 2; page 137)
18. r (Objective 3; page 142)

Completion

1. smaller, larger, luminosity (Objectives 1 and 2; pages 134–137)
2. surface temperature, hotter, cooler (Objective 3; pages 138–140)
3. mass, diameter, larger, smaller (Objectives 4 and 5; pages 143, 147, 151–153)

Self-Test

1. d (Objective 1; page 135)
2. c (Objective 1; page 135)
3. d (Objective 2; page 136; video program)
4. b (Objective 2; pages 136–137)
5. a (Objective 3; pages 140–141)
6. a (Objective 3; pages 138–140)
7. b (Objective 4; pages 147–148; video program)
8. a (Objective 4; pages 142–146; video program)
9. d (Objective 5; pages 152–153)
10. c (Objective 5; pages 149–151)

Lesson 10

Matching

1. c (Objective 1; page 157)
2. f (Objective 1; page 158)
3. k (Objective 1; page 159)
4. e (Objective 2; page 163)
5. h (Objective 1; pages 160–161)
6. g (Objective 1; pages 160–161)
7. d (Objective 2; page 162)
8. i (Objective 2; page 162)
9. l (Objective 2; page 163)
10. b (Objective 2; page 163)
11. a (Objective 1; page 158)
12. j (Objective 3; page 165)
13. n (Objective 3; page 166)
14. p (Objective 3; page 167)
15. o (Objective 3; page 167)
16. m (Objective 3; page 166)

Completion

1. gas, hydrogen, helium, dust, reddening, red, less, blue (Objective 1; pages 157–159)
2. colder, greater, hotter, greater (Objective 2; pages 162–164)
3. lesser, infrared, colder, dust, visible, gas, infrared, visible (Objective 3; pages 165–167)

Self-Test

1. a (Objective 1; page 157; video program)
2. c (Objective 1; page 158; video program)
3. d (Objective 1; page 159)
4. b (Objective 2; page 162; video program)
5. d (Objective 2; pages 162–163; video program)
6. d (Objective 2; page 164)
7. c (Objective 3; page 165; video program)
8. a (Objective 3; pages 162, 166, 167)
9. a (Objective 3; pages 166–167)
10. b (Objective 3; pages 176–179; video program)

Lesson 11

Matching

1. n (Objective 1; page 123)
2. o (Objective 1; page 123)
3. e (Objective 1; page 126)
4. f (Objective 1; page 127)
5. a (Objective 1; page 127)
6. b (Objective 1; page 165)
7. m (Objective 1; page 127)
8. c (Objective 1; page 168)
9. d (Objective 1; page 126)
10. j (Objective 3; page 169)
11. k (Objective 3; page 169)
12. h (Objective 3; page 169)
13. i (Objective 3; page 170)
14. g (Objective 3; page 171)
15. l (Objective 3; page 172)
16. p (Objective 4; page 174)
17. q (Objective 3; page 175)

Completion

1. hydrogen, helium, carbon, helium, carbon, less, more (Objective 1; pages 126–127, 165–168)
2. decrease, increase, increase, more, less, less (Objectives 4 and 5; pages 173–176)
3. increase, increase, conduction, convection, radiation, conduction, gaseous, radiation, high, convection, uniform (Objective 3; pages 169–171)
4. less, decrease (Objective 1; page 174)

Self-Test

1. d (Objective 1; page 126; video program)
2. d (Objective 1; page 126)
3. b (Objective 2; page 128; video program)
4. d (Objective 2; page 128; video program)
5. a (Objective 3; page 170)
6. c (Objective 3; page 170)
7. d (Objective 4; page 175; video program)
8. d (Objective 4; pages 175–176)
9. d (Objective 5; page 174; video program)
10. a (Objective 5; page 176; video program)

Lesson 12

Matching

1. j (Objective 1; page 186)
2. c (Objective 1; page 185)
3. n (Objective 2; page 191)
4. d (Objective 2; page 191)
5. b (Objective 2; page 194)
6. i (Objective 3; page 202)
7. k (Objective 3; page 203)
8. l (Objective 4; page 188)
9. f (Objective 5; page 195)
10. g (Objective 5; page 195)
11. a (Objective 5; page 196)
12. e (Objective 5; page 196)
13. h (Objective 5; page 197)

Completion

1. white dwarf, giant star, planetary nebula, white dwarf, nova, supernova, supergiant star, iron, supernova (Objectives 2, 3, and 5; pages 191–195, 197, 201)
2. giant star, hydrogen, helium, helium, carbon, carbon, supernova (Objectives 1 and 5; pages 183–186, 201)
3. young, old, older, younger (Objective 4; pages 188–189)

Self-Test

1. c (Objective 1; pages 183–184; video program)
2. b (Objective 1; pages 185–186, 191, 201)
3. a (Objective 2; page 190)
4. c (Objective 2; page 191; video program)
5. d (Objective 3; pages 200–201; video program)
6. a (Objective 3; page 202; video program)
7. a (Objective 4; pages 188–189)
8. d (Objective 4; pages 188–189)
9. d (Objective 5; page 197; video program)
10. d (Objective 5; pages 195–196)

Lesson 13

Matching

1. b (Objective 1; page 208)
2. c (Objective 2; page 209)
3. f (Objective 3; page 210)
4. d (Objective 3; page 217)
5. e (Objective 4; page 220)
6. a (Objective 4; page 220)
7. g (Objective 4; page 219)
8. k (Objective 4; page 220)
9. h (Objective 4; page 220)
10. i (Objective 4; page 221)
11. j (Objective 4; page 221)

Completion

1. white dwarf, neutron star, black hole, neutron star, white dwarf, black hole, white dwarf (Objectives 1, 2, 4, and 5; pages 208–210, 219, 222)
2. radio waves, neutron stars, supernovae, slower, faster, (Objective 3; pages 210–217)
3. two to three, faster, slower, slower (Objective 4; pages 219–221)

Self-Test

1. d (Objective 1; page 208)
2. c (Objective 1; pages 208–209)
3. a (Objective 2; page 209; video program)
4. c (Objective 2; page 210; video program)
5. d (Objective 3; pages 212–213)
6. d (Objective 3; page 217)
7. c (Objective 4; pages 220–221)
8. c (Objective 4; page 221)
9. b (Objective 5; page 222; video program)
10. b (Objective 5; page 222)

Lesson 14

Matching

1. j (Objective 1; page 229)
2. k (Objective 1; page 231)
3. l (Objective 1; page 231)
4. u (Objective 1; page 231)
5. v (Objective 1; page 231)
6. s (Objective 1; page 231)
7. h (Objective 1; page 232)
8. i (Objective 1; page 232)
9. w (Objective 1; page 232)
10. t (Objective 1; page 232)
11. b (Objective 2; page 235)
12. x (Objective 2; page 235)
13. e (Objective 2; page 235)
14. a (Objective 2; page 235)
15. d (Objective 2; page 236)
16. c (Objective 2; page 236)
17. m (Objective 2; page 237)
18. n (Objective 2; page 237)
19. g (Objective 2; page 238)
20. z (Objective 4; page 239)
21. y (Objective 4; page 239)
22. f (Objective 5; page 244)
23. p (Objective 5; page 246)
24. r (Objective 5; page 247)
25. q (Objective 5; page 247)
26. o (Objective 6; page 249)

Completion

1. younger, older, older, younger, older, younger, younger, older, older (Objective 4; pages 239–240)
2. grand design, density waves, flocculent, self-sustaining star formation, differential rotation (Objective 5; pages 246–248)

Self-Test

1. d (Objective 1; pages 232, 239–240; video program)
2. b (Objective 1; pages 188, 232)
3. c (Objective 1; page 231)
4. c (Objective 1; page 235)
5. a (Objective 2; page 237; video program)
6. d (Objective 2; page 238)
7. d (Objective 3; page 235; video program)
8. b (Objective 3; page 235; video program)
9. d (Objective 4; page 240)
10. c (Objective 4; pages 241–242)
11. d (Objective 5; page 244; video program)

12. c (Objective 5; pages 245–247)

13. c (Objective 6; page 249; video program)

14. c (Objective 6; page 251; video program)

Lesson 15

Matching

1. p (Objective 1; page 258)
2. q (Objective 1; page 258)
3. r (Objective 1; page 258)
4. o (Objective 1; page 259)
5. i (Objective 2; page 260)
6. c (Objective 2; page 260)
7. j (Objective 2; page 260)
8. e (Objective 2; page 262)
9. a (Objective 3; page 263)
10. b (Objective 3; page 263)
11. h (Objective 2; page 265)
12. g (Objective 2; page 265)
13. d (Objective 2; page 265)
14. n (Objective 4; page 267)
15. m (Objective 4; page 267)
16. k (Objective 4; page 272)
17. l (Objective 4; page 269)
18. f (Objective 1; page 259)

Completion

1. elliptical, E0, E7, spiral, S0, elliptical, spiral, irregular, spiral, elliptical, elliptical, S0 (Objective 1; pages 256–259)
2. distance, radial velocity, luminosity, diameter, mass, luminosity, mass (Objectives 2 and 3; pages 260–265)
3. uniformly, clusters, rich galaxy cluster, elliptical, giant, Local Group, poor galaxy cluster, irregular (Objective 4; page 267)

Self-Test

1. a (Objective 1; pages 258–259)
2. d (Objective 1; pages 258–259)
3. b (Objective 2; page 263)
4. d (Objective 2; page 260; video program)
5. a (Objective 2; pages 265–266)
6. c (Objective 3; page 263)
7. b (Objective 3; page 262; video program)
8. b (Objective 4; pages 267–268; video program)
9. d (Objective 4; page 269)
10. a (Objective 4; pages 267, 273)

Lesson 16

Matching

1. g (Objective 1; page 277)
2. e (Objective 1; page 277)
3. d (Objective 1; page 277)
4. h (Objective 1; page 278)
5. j (Objective 1; page 280)
6. i (Objective 1; page 278)
7. c (Objective 1; page 277)
8. f (Objective 1; page 283)
9. a (Objective 2; page 285)
10. b (Objective 3; page 288)
11. k (Objective 3; page 289)

Completion

1. large, red, distant, high, small, small, black hole, large, earlier, more (Objectives 2 and 4; pages 286–287, 291–292)

Self-Test

1. b (Objective 1; pages 280–282)
2. c (Objective 1; page 278; video program)
3. d (Objective 1; pages 277–278)
4. a (Objective 2; pages 285–287, 291; video program)
5. d (Objective 2; page 286; video program)
6. c (Objective 3; page 288)
7. d (Objective 3; page 288; video program)
8. b (Objective 3; pages 289–290)
9. a (Objective 4; page 292; video program)
10. a (Objective 4; pages 277, 283, 287; video program 12)

Lesson 17

Matching

1. c (Objective 1; page 296)
2. g (Objective 1; page 296)
3. l (Objective 2; page 306)
4. k (Objective 2; page 305)
5. m (Objective 3; page 300)
6. d (Objective 2; page 306)
7. j (Objective 3; page 309)

8. i (Objective 3; page 309)
9. h (Objective 3; page 309)
10. b (Objective 3; page 299)
11. e (Objective 4; page 302)
12. a (Objective 4; page 302)
13. f (Objective 5; page 304)

Completion

1. the same, largest, homogeneity, largest, the same, isotropy, big bang (Objective 2; pages 305–306)
2. finite, bounded, positive, small, infinite, unbounded, negative, large, infinite, unbounded, zero, normal, edge, center (Objective 3; page 309)

Self-Test

1. c (Objective 1; pages 298; video program)
2. c (Objective 1; page 298; video program)
3. b (Objective 2; page 305)
4. c (Objective 2; page 306)
5. a (Objective 3; page 298; video program)
6. d (Objective 3; page 309)
7. a (Objective 4; pages 301–302; video program)
8. a (Objective 4; page 302; video program)
9. d (Objective 5; page 304)
10. b (Objective 5; page 303)
11. c (Objective 6; page 317; video program 15)
12. b (Objective 6; page 318; video program 15)

Lesson 18

Matching

1. c (Objective 2; page 306)
2. e (Objective 2; page 307)
3. b (Objective 1; page 313)
4. h (Objective 3; page 316)
5. d (Objective 1; page 313)
6. a (Objective 1; page 312)
7. g (Objective 1; page 313)
8. f (Objective 4; page 311)

Completion

1. flat, open, closed, oscillating, inflationary, strong force, weak force, electromagnetic force (Objectives 1 and 2; pages 306–307, 313)

Self-Test

1. b (Objectives 1 and 2; pages 306–309; video program
2. d (Objective 1; page 313; video program)
3. b (Objectives 1 and 2; page 313; video program)
4. c (Objective 2; page 307; video program)
5. c (Objective 3; page 300; video program)
6. a (Objective 3; page 314; video program)
7. c (Objectives 3 and 4; page 307; video program)
8. c (Objective 4; page 317; video program)
9. d (Objective 4; pages 302; video program)
10. d (Objective 4; pages 311, 316; video program)

Lesson 19

Matching

1. e (Objective 3; page 324)
2. m (Objective 3; page 336)
3. f (Objective 4; pages 332–333)
4. a (Objective 4; pages 332–333)
5. v (Objective 7; page 325)
6. p (Objective 4; page 333)
7. s (Objective 5; page 330)
8. j (Objective 5; page 330)
9. i (Objective 3; page 336)
10. r (Objective 5; page 330)
11. c (Objective 5; page 330)
12. b (Objective 5; page 330)
13. h (Objective 3; page 337)
14. l (Objective 6; page 338)
15. o (Objective 6; page 339)
16. g (Objective 6; page 338)
17. u (Objective 6; page 338)
18. k (Objective 2; page 334)
19. d (Objective 6; page 340)
20. n (Objective 3; page 341)
21. q (Objective 6; page 339)
22. t (Objective 4; page 342)

Completion

1. counterclockwise, counterclockwise, Venus, Uranus, planets, minor members, counterclockwise, same (counterclockwise) (Objective 2; page 330)

    common plane, edge on, straight, horizontal, thin disk, orbits, thin disk, equators, slightly, Earth's (Objective 2; page 330)

    two, Earth, small, solid, rocky, terrestrial, large, gaseous, solid, Jupiter, Jovian, hydrogen, helium (Objective 4; pages 332–333)

    moons, asteroids, comets, rocks, ice, gas, dust, comets, gas, dust, solar nebula theory (Objective 5; pages 330–331, 334)

2. planetesimals, protoplanets, same, speeds, low, sticky coatings, electrostatic forces, planetesimals, gravity (Objective 6; page 338)

    Earthlike, heat, short-lived, radioactive, melt, high-density, crust, differentiation, original, hydrogen, helium, outgassing (Objective 6; page 339)

    temperature, high, metals, metal oxides, metallic, silicates, mantles, crusts, infalling, heat of formation (Objective 6; pages 339–340)

    molten, hydrogen, helium, outgassing, high temperatures, water, collided, volatile-rich, icy, outer, Jupiter (Objectives 3 and 4; pages 340–341)

Self Test

1. b (Objective 1; page 332)
2. c (Objective 1; page 341)
3. a (Objective 2; page 330)
4. c (Objective 2; page 330)
5. a (Objective 2; pages 324, 330, 332, 340; video program)
6. c (Objective 3; pages 341–343)
7. d (Objective 3; page 324; video program)
8. b (Objective 4; pages 332–333)
9. a (Objective 4; pages 332–333)
10. d (Objective 5; page 330)
11. d (Objective 5; page 324)
12. c (Objective 6; page 338)
13. a (Objective 6; page 338)
14. b (Objective 6; page 339)
15. c (Objective 4; page 333)
16. a (Objective 7; page 325)
17. c (Objective 7; page 327)

Lesson 20

Matching

1. i (Objective 1; page 346)
2. c (Objective 2; page 347)
3. b (Objective 2; page 347)
4. l (Objective 3; page 347)
5. h (Objective 2; page 350)
6. n (Objective 4; pages 350–351)
7. j (Objective 4; pages 350–351)
8. k (Objective 4; pages 350–351)
9. e (Objective 4; pages 350–351)
10. d (Objectives 5 and 7; pages 348–349)
11. g (Objective 5; page 348)
12. a (Objective 4; page 351)
13. m (Objective 4; pages 350–351)
14. f (Objective 4; pages 350–351)

Completion

1. convection, mantle, continents, midocean rises, midocean rifts, molten, spread, seafloor spreading, magnetism, magnetic, reversed, midocean rifts, symmetric, alternating, crust, seafloor (Objective 4; pages 348, 350–351)

2. primeval, solar nebula, hydrogen, helium, methane, ammonia, carbon dioxide, water vapor, carbon dioxide, greenhouse effect, temperature, ammonia, methane, ultraviolet, carbon dioxide, nitrogen, ozone (Objective 5; pages 348–349)

   infrared, increase, global warming, melt, rise, ozone layer, chlorofluorocarbons, ozone layer, ultraviolet radiation, increase (Objectives 5, 6, and 7; pages 349, 352)

Self-Test

1. a (Objective 1; page 346; video program)
2. c (Objective 1; pages 346–347)
3. b (Objective 2; page 347; video program)
4. a (Objective 2; pages 347–348; video program)
5. d (Objective 3; page 347; video program)
6. b (Objective 3; page 347; video program)
7. c (Objective 4; page 348; video program)
8. d (Objective 4; pages 351)
9. a (Objective 5; page 348)
10. c (Objective 5; page 348; video program)
11. c (Objective 6; pages 349, 352)
12. b (Objective 6; pages 349, 352)
13. a (Objective 7; pages 348–349; video program)
14. d (Objective 7; pages 348–349; video program)
15. b (Objective 8; page 349; video program)
16. b (Objective 8; pages 349, 352)

Lesson 21

Matching

1. d (Objective 1; page 354)
2. h (Objective 1; page 354)
3. e (Objective 2; page 353)
4. b (Objective 1; page 353)
5. j (Objective 2; page 353)
6. g (Objective 2; page 353)
7. m (Objective 1; page 353)
8. f (Objective 1; page 357)
9. c (Objective 3; page 358)
10. i (Objective 3; pages 358–359)
11. k (Objective 3; page 359)
12. a (Objective 3; page 359)
13. l (Objective 4; pages 360–361)
14. o (Objective 6; Background Notes 21A)
15. n (Objective 6; Background Notes 21A)

Completion

1. highlands, cratered, smooth, darker, seas, maria, craters, lowlands, highlands (Objective 1; pages 352–353)

   Apollo, vesicular basalt, lava, impacts, highlands, anorthosite, aluminum, calcium, lighter, lower, basalt (Objective 2; pages 352–353)

2. rapidly, broke away, iron, differentiated, fission, composition, angular momentum (Objective 3; pages 357–358)

   solar nebula, condensed, condensation, chemical, density, volatiles, evaporated, water, water (Objective 3; pages 358–359

   elsewhere, captured, capture, density, composition, slowly, capture, trajectory, orbit, tidal, trajectories, capture (Objective 3; page 359)

   large-impact, Mars, collided, mantle, moon, angular momentum, differentiated, iron, water, volatiles (Objective 3; page 359)

3. craters, maria, maria, lighter, lava, darker, lava, lobate scarps, shrinkage, interior, cooled, Mariner 10, poles, radar bright, water ice, craters (Objective 4; pages 360–361; video program)

Self-Test

1. a (Objective 1; pages 352–353; video program)
2. c (Objective 1; pages 354–355, 357; video program)
3. c (Objective 2; page 353)
4. b (Objective 2; page 357)
5. d (Objective 3; page 359)
6. a (Objective 3; page 358)
7. a (Objective 4; page 360)
8. b (Objective 4; pages 360–361)
9. c (Objective 5; page 361; video program)

10. d (Objective 5; video program)
11. a (Objective 6; Background Notes 21A and 21B)
12. b (Objective 6; video program)

Lesson 22

Matching

1. f (Objective 1; pages 364–365)
2. k (Objective 1; page 363)
3. g (Objective 1; pages 363–364)
4. b (Objective 1; page 367)
5. o (Objectives 1 and 3; page 365)
6. e (Objectives 1 and 3; video program)
7. c (Objective 3; page 365)
8. h (Objective 2; page 368)
9. a (Objective 4; page 372)
10. l (Objective 4; pages 372–373)
11. j (Objective 4; pages 371–372)
12. i (Objective 6; pages 369–370)
13. d (Objective 7; pages 376–378)

Completion

1. less, carbon dioxide, greenhouse effect, runaway greenhouse effect (Objective 2; pages 362–363)

   thick, blanket, radiation, surface, infrared, carbon dioxide, increases, high (Objective 2; pages 365–368)

2. oxygen, nitrogen, carbon dioxide, greenhouse effect, less, Venus, liquid water (Objectives 2 and 3; pages 362–363)

   carbon dioxide, water vapor, farther, lower, water vapor, liquid water, carbon dioxide, dissolve, oceans, carbon dioxide, limestone (Objectives 2 and 3; pages 365–368)

3. Mariner, Viking, 1970s, cratered, moon, craters, eroded, younger, volcanic activity (Objective 4; pages 370–372)

   lava flows, shield, Hawaii, thicker, smaller, cooled, thicker, grow, larger (Objective 4; pages 368, 370–373)

   runoff channels, outflow channels, liquid water, flooding, liquid water, ice, poles, water, thinner, pressure, water, liquid, smaller, gravity, dense (Objectives 4 and 5; pages 373–375)

Self-Test

1. d (Objective 1; page 363)
2. b (Objective 1; pages 363–365, 368; video program)
3. a (Objective 1; page 365)
4. c (Objective 1; page 365, and Figure 17-13b; video program)
5. d (Objective 2; page 362; video program)
6. c (Objective 2; pages 362–363, 368; video program)
7. b (Objective 3; pages 365–368)
8. c (Objective 3; pages 363–365; video program)
9. a (Objective 4; pages 373–375, and Figure 17-20; video program)

10. b (Objective 4; pages 371–372)

11. d (Objective 5; pages 374–375)

12. b (Objective 5; pages 374–375)

13. a (Objective 6; page 369)

14. d (Objective 6; page 369)

15. a (Objective 7; pages 376–378)

16. c (Objective 7; pages 376–378)

Lesson 23

Matching

1. c (Objective 2; page 383)
2. d (Objectives 1 and 2; page 383)
3. b (Objective 3; pages 386–387)
4. f (Objective 3; page 388)
5. a (Objective 3; page 390)
6. e (Objective 1; page 391)
7. g (Objective 7; page 395)
8. h (Objective 4; page 386)

Completion

1. Galileo, Galilean, Io, craters, craters, Voyager, volcanoes, elliptical, tidal forces, heat (Objective 3; pages 388–390)

   Europa, craters, young, rocky, liquid water, icy, Ganymede, Galilean, cratered, grooved terrain, icy, after, young, Callisto, cratered, oldest, rock, ice (Objective 3; pages 388–390)

2. Galileo, particles, gaps, Cassini, Voyager, ringlets, two, moons, gravitational, shepherding, resonances, satellites (moons), Mimas, ringlets (Objective 7; pages 394–395; video program)

Self-Test

1. a   (Objective 1; page 381; video program)
2. b   (Objective 1; page 391)
3. c   (Objective 1; page 383)
4. b   (Objective 1; page 391)
5. d   (Objective 2; page 390)
6. c   (Objective 2; page 385)
7. b   (Objective 3; page 388)
8. b   (Objective 3; page 390)
9. c   (Objective 4; page 386)
10. a   (Objective 4; page 387)
11. b   (Objective 4; pages 392, 394–395)
12. d   (Objective 5; page 382)
13. d   (Objective 5; pages 383–385; video program)
14. b   (Objective 6; page 392)
15. c   (Objective 6; page 396)
16. b   (Objective 7; page 395)
17. b   (Objective 7; page 395)
18. c   (Objective 8; video program)
19. d   (Objective 8; page 391; video program)
20. a   (Objective 2; pages 388–390)

Lesson 24

Matching

1. f (Objective 1; pages 398, 401)
2. d (Objective 4; pages 400, 403)
3. b (Objective 1; page 400)
4. c (Objective 5; page 403)
5. a (Objective 7; page 405)
6. e (Objectives 7 and 8; pages 407–408)

Matching—Planet Review

1. b (Objective 1; page 392)
2. a (Objective 1; page 385)
3. c (Objective 1; page 363)
4. d (Objective 1; pages 373–374)
5. e (Objective 1; pages 406–407)
6. a (Objective 1; page 381)
7. e (Objective 1; pages 406–407)
8. d, f (Objective 1; video program; page 375)
9. a (Objective 1; page 383)
10. f (Objective 1; video program)
11. a, b, h, i (Objective 1; pages 386, 392, 400, 405)
12. c (Objective 2; page 362)
13. none
14. g (Objective 1; page 360)
15. e (Objective 1; video program)
16. i (Objective 1; video program)
17. b (Objective 1; page 395)
18. c (Objective 1; page 362)
19. b (Objective 1; page 391)
20. e (Objective 1; page 407)

Completion

1. hydrogen, helium, metallic, liquid, blue-green, belts, bands, featureless, methane, red, yellow, blue-green (Objectives 1 and 3; pages 398–399, 401–402)

   temperatures, bland, featureless, more, blue, Great Dark Spot, heat, convection, rapid, high-speed, cyclonic (Objectives 1 and 3; pages 401–404)

2. occulted, light, ahead, symmetrical, behind, narrow, moons, particles, shepherding satellites, black, organic polymers, methane, dark, radiation belts (Objective 5; pages 400, 402–403, 405)

3. elliptical, closer, inclined (Objective 7; page 406)

   Charon, mass, eclipsing binary, diameters, one-half, ice, rocks (Objective 8; pages 407–408)

   Triton, icy, icy planetesimals, rotation, retrograde (Objective 8; page 408)

Self-Test

1. a (Objective 1; page 398)
2. d (Objective 1; page 406)
3. c (Objective 2; video program)
4. c (Objective 2; video program)
5. b (Objective 3; page 398)
6. d (Objective 3; page 401)
7. b (Objective 4; page 400)
8. a (Objective 4; video program)
9. c (Objective 5; pages 402–403)
10. b (Objective 5; page 402)
11. d (Objective 5; page 403)
12. d (Objective 6; pages 400–401)
13. c (Objective 6; pages 405–406)
14. a (Objective 7; page 405)
15. c (Objective 7; page 408)
16. a (Objective 8; pages 407–408)
17. b (Objective 8; pages 407–408)

Lesson 25

Matching

1. e (Objective 1; page 414)
2. d (Objective 1; page 414)
3. b (Objective 1; page 413)
4. c (Objective 1; page 413)
5. g (Objective 6; page 422)
6. f (Objective 3; page 414)
7. h (Objective 7; pages 423–424)
8. a (Objective 1; page 413)
9. j (Objective 7; page 424)
10. l (Objective 3; page 415, Figure 19-3a)
11. i (Objective 3; pages 412–413)
12. m (Objectives 1 and 2; page 413)
13. k (Objective 2; page 412)

Completion

1. shooting stars, meteors, meteoroids, meteorite (Objective 1; pages 412–413; video program)

    iron meteorites, iron, nickel, nitric, Widmanstätten, slowly, history (Objective 1; page 413)

    stony meteorites, chondrites, chondrules, volatiles, carbonaceous chondrites, volatiles, heating, lava, achondrites (Objective 1; pages 413–414)

    stony-iron, iron, stone, molten, rock, molten (Objective 1; page 414)

2. Mars, Jupiter, asteroids, infrared (Objective 3; pages 416–418; video program)

    bright, red, metals, chondrites, carbonaceous, carbonaceous chondrites, outer, low, heating, red, S-type, iron, nickel, iron meteorites (Objective 3; page 418)

3. tail, nucleus, snowball, less, fluffy, dark, coma, gas, dust, millions (Objective 6; pages 420–423; video program)

    gas, ionized, outward, dust, solid, dust, outward, sunlight, away (Objective 6; pages 420–423; video program)

Self-Test

1. c (Objective 1; page 413)
2. b (Objective 1; page 413)
3. c (Objective 2; pages 414–416)
4. a (Objective 2; page 415)
5. c (Objective 3; pages 412–413)
6. b (Objective 3; page 415, Figure 19-3c)
7. b (Objective 4; page 416; video program)

8. d (Objective 4; page 418)

9. b (Objective 5; page 419)

10. c (Objective 5; page 416)

11. a (Objective 6; page 423)

12. d (Objective 6; pages 420–422; video program)

13. c (Objective 7; pages 421, 424)

14. a (Objective 7; page 424)

15. b (Objective 8; Background Notes 25A)

16. c (Objective 8; pages 420–421; video program)

17. d (Objective 8; pages 420–421)

18. c (Objective 9; page 425; Background Notes 25C)

19. b (Objective 9; pages 426–427; Background Notes 25C)

Lesson 26

Matching

1. f (Objective 2; pages 432–435)
2. k (Objective 1; page 434)
3. b (Objective 2; page 435)
4. g (Objective 1; page 434)
5. m (Objective 1; page 434)
6. o (Objectives 2 and 3; page 433)
7. j (Objectives 2 and 3; page 433)
8. e (Objective 3; page 436)
9. n (Objective 3; pages 436–437)
10. h (Objective 3; page 437)
11. d (Objective 3; page 438)
12. l (Objective 3; page 439)
13. i (Objective 4; page 443)
14. c (Objective 7; page 446)
15. a (Objective 7; page 446)

Completion

1. carbon, stable, energy, cell, nucleus, twisted ladder, information, reproduces, functions (Objective 1; pages 431–433)

    deoxyribonucleic acid, phosphates, sugars, rungs, bases, four, bases, cell (Objective 2; pages 434–435)

    proteins, enzymes (Objective 2; pages 434–435)

    proteins, carbon, ribonucleic acid, information, amino acids, proteins, messenger, DNA (Objective 2; page 435)

2. stable, single, binary (Objective 4; page 443)

    billion, O, B, short, M, cool, close, tidally locked, same, constant, freeze, M, flares, close (Objective 4; page 443)

    F, G, K, cooler, long, liquid, life zone, single, cooler, F, G, K, long, life zones (Objective 4; pages 443–444)

3. distances, traveling, communication, radio waves, television, radio (Objective 7; pages 444–445)

    listening, extraterrestrial, radio waves, wavelengths, special (magic) frequencies, elements, compounds, life, neutral hydrogen, OH, water hole, hydrogen, OH, water, water, 21-centimeter, extraterrestrial, SETI (Objective 7; pages 444–447)

Self-Test

1. b (Objective 1; pages 431–433)
2. a (Objective 1; pages 432–433)
3. d (Objective 2; pages 433–435)
4. b (Objective 2; page 435)
5. c (Objective 3; page 436)

6. c (Objective 3; pages 436–437)

7. a (Objective 4; page 443; video program)

8. d (Objective 4; page 443; video program)

9. c (Objective 4; pages 441–442; Background Notes 26A)

10. a (Objective 4; page 442)

11. b (Objective 5; pages 325–329; Background Notes 26B; video program)

12. a (Objective 5; Background Notes 26B; video program)

13. d (Objective 6; pages 447–448; video program)

14. c (Objective 6; pages 447–448; video program)

15. b (Objective 7; page 446; video program)

16. a (Objective 7; page 446)